Augsburger Schriften zur Mathematik, Physik und Informatik
Band 12

herausgegeben von:
Professor Dr. F. Pukelsheim
Professor Dr. W. Reif
Professor Dr. D. Vollhardt

Bibliografische Information der Deutschen Nationalbibliothek

Die Deutsche Nationalbibliothek verzeichnet diese Publikation in der Deutschen Nationalbibliografie; detaillierte bibliografische Daten sind im Internet über http://dnb.d-nb.de abrufbar.

ISBN 978-3-8325-1969-8
ISSN 1611-4256

Logos Verlag Berlin GmbH
Comeniushof, Gubener Str. 47,
10243 Berlin
Tel.: +49 030 42 85 10 90
Fax: +49 030 42 85 10 92
INTERNET: http://www.logos-verlag.de

The Lifted Root Number Conjecture for small sets of places and an application to CM-extensions

Dissertation
zur Erlangung des akademischen Grades eines
Doktors der Naturwissenschaften,

vorgelegt der Mathematisch-Naturwissenschaftlichen Fakultät
der Universität Augsburg

von
Andreas Nickel

Februar 2008

Erstgutachter:	Prof. Dr. Jürgen Ritter
Zweitgutachter:	Prof. Dr. Philippe Cassou-Noguès
Tag der mündlichen Prüfung:	02.06.2008

Abstract

In this paper we study a famous conjecture which relates the leading terms at zero of Artin L-functions attached to a finite Galois extension L/K of number fields to natural arithmetic invariants. This conjecture is called the Lifted Root Number Conjecture (LRNC) and has been introduced by K.W. Gruenberg, J. Ritter and A. Weiss; it depends on a set S of primes of L which is supposed to be sufficiently large. We formulate a LRNC for small sets S which only need to contain the archimedean primes. We apply this to CM-extensions which we require to be (almost) tame above a fixed odd prime p. In this case the conjecture naturally decomposes into a plus and a minus part, and it is the minus part for which we prove the LRNC at p for an infinite class of relatively abelian extensions. Moreover, we show that our results are closely related to the Rubin-Stark conjecture.

Contents

Introduction **5**

1 The LRNC for small sets of places **9**

1.1 Preliminaries . . . 9
1.2 Outline of the construction . . . 15
1.3 Independence of choices . . . 19
1.4 Basic properties of Ω_ϕ . . . 26
1.5 The conjecture . . . 40

2 Tame CM-extensions **51**

2.1 Ray class groups . . . 54
2.2 L-series and Stickelberger elements . . . 56
2.3 A restatement of the LRNC on minus parts . . . 64

3 Iwasawa theory **79**

3.1 Passing to the limit . . . 79
3.2 The descent . . . 84

4 On the Rubin-Stark conjecture **91**

4.1 The conjecture . . . 91
4.2 The tamely ramified case . . . 93

A Removing $\mu_- = 0$ **97**

Introduction

In this paper we study a famous conjecture which relates the leading terms at zero of Artin L-functions attached to a finite Galois extension L/K of number fields to natural arithmetic invariants. This conjecture is called the Lifted Root Number Conjecture (LRNC) and has been introduced by K.W. Gruenberg, J. Ritter and A. Weiss [GRW]; it depends on a set S of primes of L which is supposed to be sufficiently large. We formulate a LRNC for small sets S which only need to contain the archimedean primes. We apply this to CM-extensions which we require to be (almost) tame above a fixed odd prime p. In this case the conjecture naturally decomposes into a plus and a minus part, and it is the minus part for which we prove the LRNC at p for an infinite class of relatively abelian extensions. Moreover, we show that our results are closely related to the Rubin-Stark conjecture.

Some history

Let L/K be a finite Galois extension of number fields with Galois group G. T. Chinburg [Ch1] defined an algebraic invariant $\Omega(L/K)$ for the extension L/K. He conjectured that $\Omega(L/K)$, which is an element in $K_0(\mathbb{Z}G)$, equals the root number class $W(L/K)$, an analytic invariant defined by Ph. Cassou-Noguès and A. Fröhlich in terms of Artin root numbers. In [Ch2] he introduced two further algebraic invariants in $K_0(\mathbb{Z}G)$, called $\Omega_i(L/K)$, $i = 1, 2, 3$, where $\Omega_3(L/K) = \Omega(L/K)$. These invariants are related by the equation

$$\Omega_2(L/K) = \Omega_1(L/K) \cdot \Omega_3(L/K).$$

Chinburg conjectured that $\Omega_1(L/K) = 1$, and hence that $\Omega_2(L/K)$ also equals the root number class. In addition, he proved the Ω_2-conjecture for at most tamely ramified extensions.

All these conjectures have meanwhile been lifted to corresponding conjectures in $K_0T(\mathbb{Z}G)$; so the LRNC is Chinburg's Ω_3-conjecture in $K_0T(\mathbb{Z}G)$ rather than in $K_0(\mathbb{Z}G)$, whereas the conjectures in [BB] and [BrB] are the same concerning Chinburg's Ω_2 and Ω_1-conjecture, respectively. The LRNC assumes the validity of Stark's conjecture which guarantees the Galois compatibility of a certain homomorphism on the characters of G. D. Burns [B1] defined an element $T\Omega(L/K, 0) \in K_0(\mathbb{Z}G, \mathbb{R})$ which lies in $K_0(\mathbb{Z}G, \mathbb{Q})$ if and only if Stark's conjecture is true. He also showed in loc.cit. that $T\Omega(L/K,0)$ vanishes if and only if the LRNC holds, and that the LRNC is equivalent to the Equivariant Tamagawa Number Conjecture for the pair $(h^0(\mathrm{Spec}(L))(0), \mathbb{Z}G)$. In [B3] he has shown that this conjecture implies a whole family of related conjectures as the Rubin-Stark conjecture and the refined class number formulas of Gross, Tate and Aoki, Lee and Tan.

The LRNC is known to be true for abelian extensions $L/\mathbb{Q}$ as proved by D. Burns and C. Greither [BG1] with the exclusion of the 2-primary part; M. Flach [Fl] extended the argument to cover the 2-primary part as well. If L is in addition totally real, the LRNC was independently proved in [RW3, RW4]. Some relatively abelian results are due to W. Bley [Bl]. He showed that if L/K is a finite abelian extension, where K is an imaginary quadratic field which has class number one, then the LRNC holds for all intermediate extensions L/F such that $[L : F]$ is odd and divisible only by primes which split completely in $K/\mathbb{Q}$.

Outline of the thesis

In the first chapter we give a reformulation of the LRNC for small sets of places S. If L/K is an abelian CM-extension and one restricts to minus parts, this has recently been done by C. Greither [Gr3], where the author is interested in computing the Fitting ideal of the Pontryagin dual of minus class groups via the LRNC.
The algebraic objects of the LRNC are invariants $\Omega_\phi \in K_0T(\mathbb{Z}G)$ depending on equivariant maps ϕ. All these Ω_ϕ are mapped to Chinburg's $\Omega_3(L/K)$ via the natural connecting homomorphism $K_0T(\mathbb{Z}G) \to K_0(\mathbb{Z}G)$. Let S be a set of places of L which is large in the sense that it contains all the infinite primes, all primes which ramify in L/K and enough primes to generate the ideal class group of L. J. Tate [Ta1] constructed a canonical element τ in $\mathrm{Ext}^2_G(\Delta S, E_S)$, where ΔS is the kernel of the augmentation map $\mathbb{Z}S \twoheadrightarrow \mathbb{Z}$, and E_S denotes the S-units in L. A sequence

$$E_S \rightarrowtail A \to B \twoheadrightarrow \Delta S,$$

whose extension class is τ, and where A and B are cohomologically trivial G-modules, is called a Tate-sequence. The main objects occurring in the definition of Ω_ϕ are a Tate-sequence and an injection

$$\phi : \Delta S \rightarrowtail E_S.$$

The LRNC now asserts that Ω_ϕ is represented by the homomorphism

$$\chi \mapsto A_\phi(\check{\chi}) W(L/K, \check{\chi}),$$

where $\check{\chi}$ denotes the contragredient of a character χ of G, A_ϕ is the quotient of the Stark-Tate regulator and the leading term at zero of the S-truncated

Artin L-function attached to χ, and $W(L/K, \chi)$ is defined in terms of Artin root numbers.
If S is not large, but still contains the infinite primes, J. Ritter and A. Weiss [RW1] constructed a Tate-sequence

$$E_S \rightarrowtail A \to B \twoheadrightarrow \nabla$$

with an explicitly determined G-module ∇. But in general there do not exist injections $\nabla \rightarrowtail E_S$. After a few preliminaries we show how to remedy this problem and give a definition of Ω_ϕ for small sets S. We prove that the definition is independent of all the choices made during the construction (apart from ϕ and S), and hence we can view Ω_ϕ as an arithmetic invariant of L/K. Then we discuss how Ω_ϕ varies if we change ϕ or enlarge the set S. This leads us to the definition of a modified Stark-Tate regulator and a conjectural representing homomorphism of Ω_ϕ. We call this the LRNC for small sets of places; of course, it is equivalent to the LRNC for large sets of places.

In the second chapter we apply this reformulation to CM-extensions which are assumed to be tame above a fixed odd prime p. Actually, we permit a slightly more general class of extensions. The primary idea was to restrict ourselves to minus parts and to use the LRNC for the set S_∞ of all infinite primes. In this case, the leftmost term of the corresponding Tate-sequence consists just of the roots of unity in L which seems easy to handle. The rightmost term, however, is no longer torsion free and thus becomes more complicated. For this reason we have to choose a set of places for which both sides are comfortable to some degree. This turns out to be a set which contains only totally decomposed (and thus unramified) primes.
In the first section of this chapter, we prove that the p-part of a certain ray class group of L is cohomologically trivial on minus parts. We give a definition of non-abelian Stickelberger elements in section two. These elements can be viewed as representing homomorphisms of elements in $K_0T(\mathbb{Z}G)$. In the last section, we show that the minus part of the LRNC at p holds if and only if the ray class groups treated in section one are represented by corresponding Stickelberger elements.

Note that taking minus parts simplifies matters for various reasons. First, Stark's conjecture is known to be true for odd characters. Moreover, the infinite primes consist of pairs of complex conjugate embeddings and hence neatly drop out on minus parts, i.e. $(\mathbb{Z}S_\infty)^- = 0$. At last, when Iwasawa theory comes into play in chapter three, taking minus parts provides an opportunity of an easier descent.

In chapter three, we assume the Galois group G to be abelian. In this case one can translate the minus part of the LRNC at p to the assertion that the Fitting ideal of the above ray class group is generated by the corresponding Stickelberger element. We pass to the limit and get the respective statement

at infinite level thanks to a result of C. Greither [Gr2] provided that the Iwasawa μ-invariant vanishes. We will remove this hypothesis for a special class of extensions (including the case $p \nmid |G|$) in the appendix. Note that the vanishing of μ is a long standing conjecture; the most general result is still due to B. Ferrero and L. Washington [FW] and says that $\mu = 0$ for absolute abelian extensions.
For the descent we use a method which is due to A. Wiles [Wi2] in the extended version by C. Greither [Gr1]. For this, we have to assume a slightly more restrictive hypothesis on the primes above p.

The exclusion of the prime $p = 2$ has two main reasons; the Iwasawa main conjecture is not known in this case, and taking minus parts is not exact if 2 is not invertible in the ground ring.

In the last chapter we prove the Rubin-Stark conjecture for the same class of extensions. The main ingredient is a result of C. Popescu [P3]. He proved that the Rubin-Stark conjecture follows from the stronger statement that the Fitting ideal of a certain ray class group of L contains a particular Stickelberger element. These are not the same ray class groups resp. Stickelberger elements as in the previous chapters, but they are related to them closely enough.
As already mentioned above, D. Burns [B3] has shown that the LRNC always implies the Rubin-Stark conjecture. Thus, we have reproved this result for (almost) tame extensions. Our approach uses more explicit methods and we indeed prove a stronger result which is called the Strong Brumer-Stark conjecture in [P3]. But note that this conjecture does not hold in general, as one can see from the results in [GK].

Acknowledgements

It is a pleasure to thank my supervisor Jürgen Ritter for his great support. His inspiring teaching and the illuminating discussions with him have made my studies before and during my PhD instructive and enjoyable.

I am grateful to Werner Bley, whose lectures I have attended, for his encouragement and guidance. I initially shared my office with Simone Schuierer and subsequently with Irene Sommer. I am indebted to them for many discussions and the pleasant atmosphere at work.

I also wish to thank my wife and my parents for their non-mathematical, but just as important, support.

Chapter 1

The Lifted Root Number Conjecture for small sets of places

1.1 Preliminaries

Duals

Let G be a finite group. For each $\mathbb{Z}G$-module M we write M^0 for its $\mathbb{Z}$-dual $\mathrm{Hom}_{\mathbb{Z}}(M,\mathbb{Z})$ with the G-action formula $(gf)(m) = gf(g^{-1}m) = f(g^{-1}m)$ for $g \in G$, $f \in M^0$ and $m \in M$. Note that there is a natural $\mathbb{Z}G$-isomorphism $\mathbb{Z}G \simeq \mathbb{Z}G^0$ that sends each $g \in G$ to the homomorphism $h \mapsto \delta_{gh}$. Of course, the δ on the righthand side is Kronecker's.
Under this identification, the dual of the natural augmentation map $\mathbb{Z}G \twoheadrightarrow \mathbb{Z}$ is the map $\mathbb{Z} \rightarrowtail \mathbb{Z}G$ that sends 1 to $N_G = \sum_{g\in G} g$. Thus, we get a $\mathbb{Z}G$-isomorphism

$$\Delta G^0 \simeq \mathbb{Z}G/N_G, \qquad (1.1)$$

where ΔG denotes the kernel of the augmentation map.

Sections

Let R be a (not necessarily commutative) ring with 1. Consider the following commutative diagram of R-modules with exact rows:

$$\begin{array}{ccccc} M_1' & \overset{\iota_1}{\hookrightarrow} & M_1 & \overset{\pi_1}{\twoheadrightarrow} & M_1'' \\ \downarrow g' & & \downarrow g & & \downarrow g'' \\ M_2' & \overset{\iota_2}{\hookrightarrow} & M_2 & \overset{\pi_2}{\twoheadrightarrow} & M_2'' \end{array} \qquad (1.2)$$

Definition 1.1.1 *Two R-homomorphisms $\tau_1 : M_1'' \to M_1$ and $\tau_2 : M_2'' \to M_2$ are called* ***commutative sections*** *if $\pi_i \circ \tau_i = \mathrm{id}$ for $i = 1, 2$ (i.e. both τ_i are sections) and $g\tau_1 = \tau_2 g''$.*

We will also refer to R-homomorphisms $\sigma_1 : M_1 \to M_1'$ *and* $\sigma_2 : M_2 \to M_2'$ *as commutative sections if* $\sigma_i \iota_i = \mathrm{id}$ *for* $i = 1, 2$ *and* $g'\sigma_1 = \sigma_2 g$.

Lemma 1.1.2 *Keep the notation of diagram (1.2) and the above definition.*

(1) *There are commutative sections* τ_1 *and* τ_2 *if and only if there are commutative sections* σ_1 *and* σ_2.

(2) *Assume that the maps* g', g, g'' *are injective and that* R *is a semisimple* K*-algebra over a field* K. *Then there always exist commutative sections* τ_1 *and* τ_2.

PROOF.

(1) If τ_1 and τ_2 are commutative sections, define $\sigma_i = \mathrm{id} - \tau_i \pi_i$ for $i = 1, 2$. It is easy to verify that σ_1 and σ_2 are commutative sections. Conversely, if σ_1 and σ_2 are commutative sections, define $\tau_i(m_i'') = m_i - \sigma_i(m_i)$ for $i = 1, 2$, where $m_i'' \in M_i''$ and m_i is any preimage of m_i'' in M_i. Again, it is easy to see that the definition is independent of the choice of m_i and that τ_1 and τ_2 in fact are commutative sections.

(2) This is Lemma 1.4 in [B2]. □

K-theory

Let R be a left noetherian ring with 1 and $\mathrm{PMod}(R)$ the category of all finitely generated projective R-modules. We write $K_0(R)$ for the Grothendieck group of $\mathrm{PMod}(R)$, and $K_1(R)$ for the Whitehead group of R which is the abelianized infinite general linear group. If S is a multiplicatively closed subset of the center of R which contains no zero divisors, $1 \in S$, $0 \notin S$, we denote the Grothendieck group of $\mathrm{T_S Mod}(R)$, the category of all S-torsion R-modules of finite projective dimension, by $K_0 S(R)$. Writing R_S for the ring of quotients of R with denominators in S we have the Localization Sequence (cf. [CR2], p. 65)

$$K_1(R) \to K_1(R_S) \xrightarrow{\partial} K_0 S(R) \to K_0(R) \to K_0(R_S). \qquad (1.3)$$

If T is a ring that contains R and M is an R-module, we will often write TM instead of $T \otimes_R M$. Moreover, if G is a group and $M = \Delta G$ is the kernel of the augmentation map $RG \twoheadrightarrow R$, we set $\Delta_T G := T \otimes_R \Delta G$. In the case $R = \mathbb{Z}$, $T = \mathbb{Z}_p$ for a prime p, we write $\Delta_p G$ instead of $\Delta_{\mathbb{Z}_p} G$.

Specializing to group rings $\mathbb{Z}G$ for finite groups G and $S = \mathbb{Z} \backslash \{0\}$ we write $K_0 T(\mathbb{Z}G)$ instead of $K_0 S(\mathbb{Z}G)$. So (1.3) reads

$$K_1(\mathbb{Z}G) \to K_1(\mathbb{Q}G) \xrightarrow{\partial} K_0 T(\mathbb{Z}G) \to K_0(\mathbb{Z}G) \to K_0(\mathbb{Q}G). \qquad (1.4)$$

Note that a finitely generated $\mathbb{Z}G$-module has finite projective dimension if and only if it is a G-c.t. (short for cohomologically trivial) module. Indeed, the projective dimension is lower or equal to 1 in this case. Further, recall that the relative K-group $K_0(\mathbb{Z}G, \mathbb{Q})$ is generated by elements of the form (P_1, ϕ, P_2) with finitely generated projective modules P_1 and P_2 and a $\mathbb{Q}G$-isomorphism $\phi : \mathbb{Q}P_1 \to \mathbb{Q}P_2$, and that there is an isomorphism

$$i_G : K_0T(\mathbb{Z}G) \simeq K_0(\mathbb{Z}G, \mathbb{Q}). \tag{1.5}$$

If a c.t. torsion $\mathbb{Z}G$-module T has projective resolution $P_1 \overset{\iota}{\rightarrowtail} P_0 \twoheadrightarrow T$, this isomorphism sends the corresponding element $[T] \in K_0T(\mathbb{Z}G)$ to $(P_1, \mathbb{Q} \otimes \iota, P_0) \in K_0(\mathbb{Z}G, \mathbb{Q})$.
We also shortly explain the map $i_G \circ \partial$. Any element of $K_1(\mathbb{Q}G)$ can be written in the form $[\mathbb{Q}G^n, \phi]$, where $n \in \mathbb{N}$ and ϕ is a $\mathbb{Q}G$-automorphism of $\mathbb{Q}G^n$. Then $i_G(\partial([\mathbb{Q}G^n, \phi])) = (\mathbb{Z}G^n, \phi, \mathbb{Z}G^n)$.
If p is a finite rational prime, the local analogue of sequence (1.4) is

$$K_1(\mathbb{Z}_pG) \to K_1(\mathbb{Q}_pG) \xrightarrow{\partial_p} K_0T(\mathbb{Z}_pG) \to 0, \tag{1.6}$$

and we have an isomorphism

$$K_0T(\mathbb{Z}G) \simeq \bigoplus_{p \nmid \infty} K_0T(\mathbb{Z}_pG). \tag{1.7}$$

For later use, we state the following K_1-Simplification Lemma which is taken from [GRW], p.50:

Lemma 1.1.3 *Suppose that we have given a diagram of $\mathbb{Q}G$-modules*

$$\begin{array}{ccccc} M_1' & \hookrightarrow & M_1 & \twoheadrightarrow & M_1'' \\ \simeq \downarrow g' & & & & \simeq \downarrow g'' \\ M_2' & \hookrightarrow & M_2 & \twoheadrightarrow & M_2'' \end{array}$$

and $\mathbb{Q}G$-isomorphisms $g, h : M_1 \to M_2$ each of which makes the diagram commutative.
For any $\mathbb{Q}G$-isomorphism $\gamma : M_2 \to M_1$ we then have equalities

$$[M_1, \gamma g] = [M_1, \gamma h],$$

$$[M_2, g\gamma] = [M_2, h\gamma]$$

in $K_1(\mathbb{Q}G)$.

To give a convenient formulation of the LRNC for small sets of places, we need to define elements $(A, \phi, B) \in K_0(\mathbb{Z}G, \mathbb{Q})$, where A is a finitely generated c.t. $\mathbb{Z}G$-module, B is $\mathbb{Z}G$-projective and $\phi : \mathbb{Q}A \to \mathbb{Q}B$ is a $\mathbb{Q}G$-isomorphism.

Definition 1.1.4 *Let A be a finitely generated c.t. $\mathbb{Z}G$-module, B projective and $\phi : \mathbb{Q}A \to \mathbb{Q}B$ a $\mathbb{Q}G$-isomorphism.*
Choose a projective resolution $P_1 \rightarrowtail P_0 \twoheadrightarrow A$ of A and an isomorphism ϕ_0 making the following diagram commutative:

$$\begin{array}{ccccc} \mathbb{Q}P_1 & \hookrightarrow & \mathbb{Q}P_0 & \twoheadrightarrow & \mathbb{Q}A \\ \| & & \downarrow{\scriptstyle \phi_0} & & \downarrow{\scriptstyle \phi} \\ \mathbb{Q}P_1 & \hookrightarrow & \mathbb{Q}(P_1 \oplus B) & \twoheadrightarrow & \mathbb{Q}B \end{array}$$

Here, the lower sequence is the canonical one. Then we define:

$$(A, \phi, B) = -(B, \phi^{-1}, A) := (P_0, \phi_0, P_1 \oplus B) \in K_0(\mathbb{Z}G, \mathbb{Q}).$$

Of course, we have to check the following:

Lemma 1.1.5 *(A, ϕ, B) is well defined.*

PROOF.[1] Taking another isomorphism $\tilde{\phi}_0 : \mathbb{Q}P_0 \to \mathbb{Q}(P_1 \oplus B)$ yields a commutative diagram

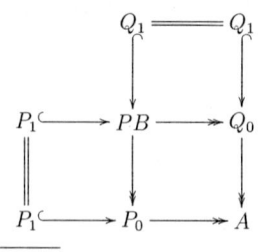

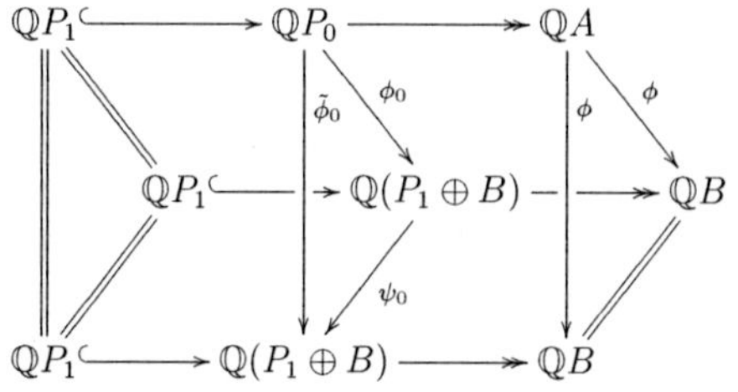

which defines an isomorphism ψ_0. Hence, we find that

$$(P_0, \tilde{\phi}_0, P_1 \oplus B) - (P_0, \phi_0, P_1 \oplus B) = (P_1 \oplus B, \psi_0, P_1 \oplus B) = 0$$

in $K_0(\mathbb{Z}G, \mathbb{Q})$. Thus, (A, ϕ, B) is independent of the choice of ϕ_0.
If we choose a second projective resolution $Q_1 \rightarrowtail Q_0 \twoheadrightarrow A$, we define PB to be the pull-back of the two surjections onto A; thus

$$\begin{array}{ccccc} & & Q_1 & = & Q_1 \\ & & \downarrow & & \downarrow \\ P_1 & \hookrightarrow & PB & \twoheadrightarrow & Q_0 \\ \| & & \downarrow & & \downarrow \\ P_1 & \hookrightarrow & P_0 & \twoheadrightarrow & A \end{array}$$

[1] In terms of Euler characteristics we have an equality $(A, \phi, B) = \chi_{\mathbb{Z}G,\mathbb{Q}G}(C^{\cdot}, \phi^{-1})$, where $C^{\cdot}$ is the perfect complex $\ldots \to 0 \to A \to B \to 0 \to \ldots$, where the position of A is in degree zero and all maps are zero. Hence, one can alternatively use the results in [B2] to show that (A, ϕ, B) is well defined.

We obtain an exact sequence $P_1 \oplus Q_1 \rightarrowtail PB \twoheadrightarrow A$, which again is a projective resolution of A. Hence, we obtain the front and back faces of the following diagram:

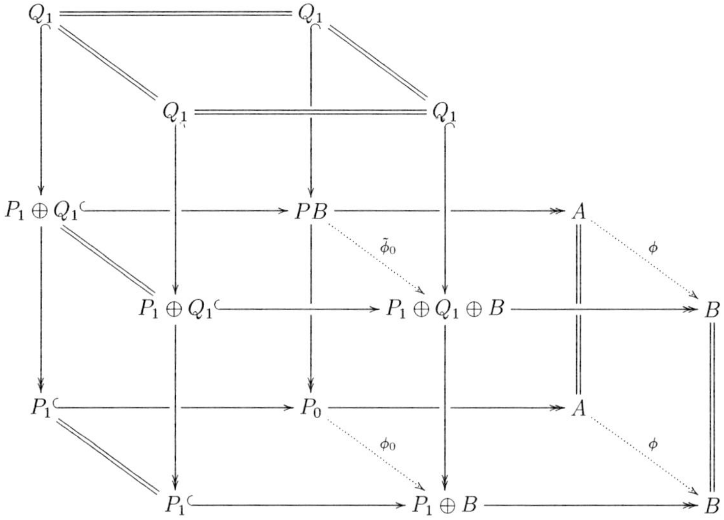

The dotted maps only exist after tensoring with $\mathbb{Q}$. Here, the isomorphism ϕ is given; the isomorphism $\tilde{\phi}_0$ is chosen to make the upper part of the diagram commute, and then $\tilde{\phi}_0$ induces the isomorphism ϕ_0.
We find that $(P_0, \phi_0, P_1 \oplus B)$ equals $(PB, \tilde{\phi}_0, P_1 \oplus Q_1 \oplus B)$ and therefore it equals $(Q_0, \psi_0, Q_1 \oplus B)$ by symmetry, where ψ_0 is constructed in exactly the same way as ϕ_0. □

We can calculate with the triples (A, ϕ, B) as usual:

Lemma 1.1.6 *Let A, A', A'' be finitely generated c.t. $\mathbb{Z}G$-modules and B, B', B'' projective $\mathbb{Z}G$-modules.*

(1) If $\phi : \mathbb{Q}A \to \mathbb{Q}B$ and $\psi : \mathbb{Q}B \to \mathbb{Q}B'$ are $\mathbb{Q}G$-isomorphisms, then

$$(A, \psi\phi, B') = (A, \phi, B) + (B, \psi, B').$$

(2) If $\phi : \mathbb{Q}B \to \mathbb{Q}A$ and $\psi : \mathbb{Q}A \to \mathbb{Q}B'$ are $\mathbb{Q}G$-isomorphisms, then

$$(B, \psi\phi, B') = (B, \phi, A) + (A, \psi, B').$$

(3) If $A' \rightarrowtail A \twoheadrightarrow A''$ and $B' \rightarrowtail B \twoheadrightarrow B''$ are exact sequences of $\mathbb{Z}G$-modules

and ϕ', ϕ, ϕ'' are $\mathbb{Q}G$-isomorphisms such that the diagram

$$\begin{array}{ccccc} \mathbb{Q}A' & \hookrightarrow & \mathbb{Q}A & \twoheadrightarrow & \mathbb{Q}A'' \\ \downarrow{\scriptstyle \phi'} & & \downarrow{\scriptstyle \phi} & & \downarrow{\scriptstyle \phi''} \\ \mathbb{Q}B' & \hookrightarrow & \mathbb{Q}B & \twoheadrightarrow & \mathbb{Q}B'' \end{array}$$

commutes, then

$$(A, \phi, B) = (A', \phi', B') + (A'', \phi'', B'').$$

PROOF. (i) and (ii) directly follow from the definition and the corresponding rules in $K_0(\mathbb{Z}G, \mathbb{Q})$. For (iii) we construct the diagram

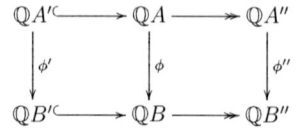

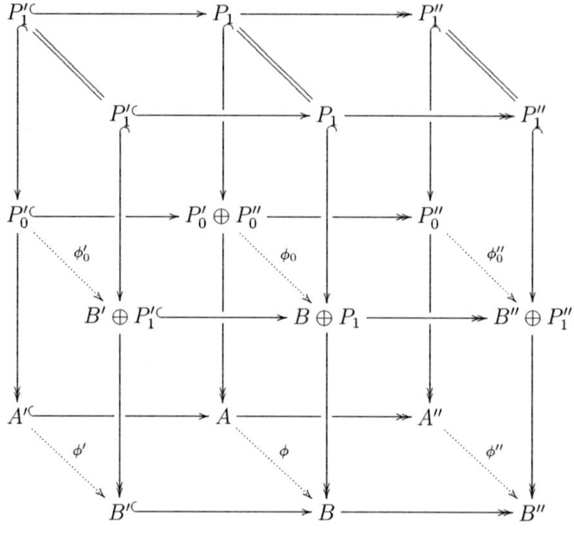

Here, we choose projective resolutions of A' and A'' which determine a projective resolution of A by the Horseshoe Lemma. Again, the dotted maps only exist after tensoring with $\mathbb{Q}$. We first choose the isomorphism ϕ_0 which induces appropriate isomorphisms ϕ_0' and ϕ_0''. The assertion is now easily read off the diagram[2]. □

[2]Alternatively, one can again trace back the above properties to the corresponding properties of refined Euler characteristics.

REMARK.

(1) If A is a c.t. torsion $\mathbb{Z}G$-module, then

$$\begin{array}{ccccc} \mathbb{Q}P_1 & \stackrel{\iota}{\hookrightarrow} & \mathbb{Q}P_0 & \twoheadrightarrow & \mathbb{Q}A \\ \| & & \downarrow{\scriptstyle \iota^{-1}} & & \downarrow{\scriptstyle \simeq} \\ \mathbb{Q}P_1 & = & \mathbb{Q}P_1 & \longrightarrow & 0 \end{array}$$

shows that $i_G([A]) = -(A,0,0) = (0,0,A)$ in $K_0(\mathbb{Z}G, \mathbb{Q})$.

(2) We can replace $K_0(\mathbb{Z}G, \mathbb{Q})$ by $K_0(\mathbb{Z}_pG, \mathbb{Q}_p)$ for any prime p. Everything remains the same except for the obvious modifications.

Hom description

Let G be a finite group, p a finite rational prime and $R(G)$ (resp. $R_p(G)$) the ring of virtual characters of G with values in $\mathbb{Q}^c$ (resp. $\mathbb{Q}_p^c$), an algebraic closure of $\mathbb{Q}$ (resp. $\mathbb{Q}_p$). Choose a number field F, Galois over $\mathbb{Q}$ with Galois group Γ, which is large enough such that all representations of G can be realized over F. Let $\wp$ be a prime of F above p. Then there is an isomorphism (for this and the following cf. [GRW], Appendix A)

$$\begin{array}{rcl} \mathrm{Det} : K_1(\mathbb{Q}_pG) & \stackrel{\simeq}{\longrightarrow} & \mathrm{Hom}_{\Gamma_\wp}(R_p(G), F_\wp^\times) \\ {[X, g]} & \mapsto & [\chi \mapsto \det(g|\mathrm{Hom}_{F_\wp G}(V_\chi, F_\wp \otimes_{\mathbb{Q}_p} X))], \end{array}$$

where V_χ is a $F_\wp G$-module with character χ. Combined with the localization sequence (1.6) this gives the local Hom description

$$K_0T(\mathbb{Z}_pG) \simeq \mathrm{Hom}_{\Gamma_\wp}(R_p(G), F_\wp^\times)/\mathrm{Det}\,(\mathbb{Z}_pG^\times). \tag{1.8}$$

One globally has

$$K_0T(\mathbb{Z}G) \simeq \mathrm{Hom}_\Gamma^+(R(G), J_F)/\mathrm{Det}\,U(\mathbb{Z}G), \tag{1.9}$$

where J_F denotes the idèle group of F and $U(\mathbb{Z}G)$ the unit idèles of $\mathbb{Z}G$. The $+$ indicates that a homomorphism $\phi \in \mathrm{Hom}_\Gamma^+(R(G), J_F)$ takes values in $\mathbb{R}^+$ for symplectic characters.

1.2 Outline of the construction

Let L/K be a finite Galois extension of number fields with Galois group G and S a finite G-invariant set of places of L which contains the set S_∞ of all the archimedean primes. In [RW1] the authors derive an exact sequence of finitely generated $\mathbb{Z}G$-modules

$$E_S \rightarrowtail A \to B \twoheadrightarrow \nabla, \tag{1.10}$$

which has a uniquely determined extension class in $\mathrm{Ext}^2_G(\nabla, E_S)$. Note that the sequence itself is not unique. We will refer to a sequence (1.10) as a Tate-sequence for S. Here, E_S is the group of S-units of L, A is c.t., B projective and ∇ fits into an exact sequence of G-modules

$$\mathrm{cl}_S \rightarrowtail \nabla \twoheadrightarrow \overline{\nabla}.$$

Indeed, the S-class group of L is the torsion submodule of ∇, hence $\overline{\nabla}$ is a $\mathbb{Z}G$-lattice. To give a description of $\overline{\nabla}$, we have to introduce some further notation:
For a prime $\mathfrak{P}$ of L we write $\mathfrak{p} = \mathfrak{P} \cap K$ for the prime below $\mathfrak{P}$, $G_{\mathfrak{P}}$ for the decomposition group attached to $\mathfrak{P}$ and $I_{\mathfrak{P}}$ for the inertia subgroup. We denote the Frobenius generator of the Galois group $\overline{G_{\mathfrak{P}}} = G_{\mathfrak{P}}/I_{\mathfrak{P}}$ of the corresponding residue field extension by $\phi_{\mathfrak{P}}$.

The inertial lattice of the local extension $L_{\mathfrak{P}}/K_{\mathfrak{p}}$ is defined to be the $\mathbb{Z}G_{\mathfrak{P}}$-lattice (cf. [GW] or [We] p. 42)

$$W_{\mathfrak{P}} = \{(x, y) \in \Delta G_{\mathfrak{P}} \oplus \mathbb{Z}\overline{G_{\mathfrak{P}}} : \overline{x} = (\phi_{\mathfrak{P}} - 1)y\}, \tag{1.11}$$

where $\Delta G_{\mathfrak{P}}$ is the kernel of the augmentation map $\mathbb{Z}G_{\mathfrak{P}} \to \mathbb{Z}$. Note that $W_{\mathfrak{P}} \simeq \mathbb{Z}G_{\mathfrak{P}}$ if the local extension $L_{\mathfrak{P}}/K_{\mathfrak{p}}$ is unramified. Projecting on the first component yields an exact sequence of $G_{\mathfrak{P}}$-modules

$$\mathbb{Z} \rightarrowtail W_{\mathfrak{P}} \twoheadrightarrow \Delta G_{\mathfrak{P}}. \tag{1.12}$$

The $\mathbb{Z}$-dual of this sequence induces a surjection $W^0_{\mathfrak{P}} \twoheadrightarrow \mathbb{Z}^0 = \mathbb{Z}$. If we combine these surjections and the augmentation map $\mathbb{Z}S \twoheadrightarrow \mathbb{Z}$, we get an exact sequence

$$\overline{\nabla} \rightarrowtail \mathbb{Z}S \oplus \bigoplus_{\mathfrak{P} \in S^*_{\mathrm{ram}} \setminus (S \cap S_{\mathrm{ram}})^*} \mathrm{ind}\,^G_{G_{\mathfrak{P}}}(W^0_{\mathfrak{P}}) \twoheadrightarrow \mathbb{Z} \tag{1.13}$$

where the sum runs over a fixed set of representatives of all ramified primes which are not in S, one for each orbit of the action of G on the primes of L. Due to this characterization of $\overline{\nabla}$ we have

Lemma 1.2.1 *Let L/K be a finite Galois extension of number fields with Galois group G and S a finite G-invariant set of places of L which contains all the archimedean primes. Moreover, let $\overline{\nabla}$ be as in (1.13) and C a free $\mathbb{Z}G$-module of rank $|S^*_{\mathrm{ram}} \setminus (S \cap S_{\mathrm{ram}})^*|$.*
Then there exist $\mathbb{Q}G$-isomorphisms $\mathbb{Q}\overline{\nabla} \xrightarrow{\simeq} \mathbb{Q}(E_S \oplus C)$.

PROOF. We have the following commutative diagram:

$$\begin{array}{ccccc}
\Delta S & \hookrightarrow & \mathbb{Z}S & \xrightarrow{\text{aug}} & \mathbb{Z} \\
\downarrow & & \downarrow & & \| \\
\overline{\nabla} & \hookrightarrow & \mathbb{Z}S \oplus \bigoplus \operatorname{ind}_{G_{\mathfrak{P}}}^{G}(W_{\mathfrak{P}}^{0}) & \twoheadrightarrow & \mathbb{Z} \\
\downarrow & & \downarrow & & \\
\bigoplus \operatorname{ind}_{G_{\mathfrak{P}}}^{G}(W_{\mathfrak{P}}^{0}) & = & \bigoplus \operatorname{ind}_{G_{\mathfrak{P}}}^{G}(W_{\mathfrak{P}}^{0}) & &
\end{array}$$

where all direct sums are taken over the primes $\mathfrak{P} \in S_{\text{ram}}^{*} \setminus (S \cap S_{\text{ram}})^{*}$, and where the middle sequence is (1.13). The left column of the diagram gives an isomorphism $\mathbb{Q}\overline{\nabla} \simeq \mathbb{Q}(\Delta S \oplus \bigoplus \operatorname{ind}_{G_{\mathfrak{P}}}^{G}(W_{\mathfrak{P}}^{0}))$. Since the Dirichlet map

$$\begin{array}{rcl}
\lambda_S : E_S & \longrightarrow & \Delta_{\mathbb{R}} S \\
e & \mapsto & -\sum_{\mathfrak{P} \in S} \log |e|_{\mathfrak{P}} \mathfrak{P}
\end{array} \tag{1.14}$$

induces an $\mathbb{R}G$-isomorphism $\mathbb{R} \otimes E_S \rightarrow \Delta_{\mathbb{R}} S$, there also exist $\mathbb{Q}G$-isomorphisms $\Delta_{\mathbb{Q}} S \rightarrow \mathbb{Q}E_S$ by the Noether-Deuring Theorem. Finally, (1.12) shows that $\mathbb{Q}G \simeq \mathbb{Q}\operatorname{ind}_{G_{\mathfrak{P}}}^{G}(W_{\mathfrak{P}}) \simeq \mathbb{Q}\operatorname{ind}_{G_{\mathfrak{P}}}^{G}(W_{\mathfrak{P}}^{0})$. □

In order to get an element $\Omega_\phi \in K_0(\mathbb{Z}G, \mathbb{Q})$ analogously to the Ω_ϕ of [GRW], we split sequence (1.10) into two parts:

$$E_S \rightarrowtail A \twoheadrightarrow W \text{ and } W \rightarrowtail B \twoheadrightarrow \nabla \tag{1.15}$$

We will refer to it as the left and the right part of the Tate-sequence. From the construction of the Tate-sequence for small sets S one gets the following diagram, which we can take for a definition of the $\mathbb{Z}G$-lattice R:

$$\begin{array}{ccccc}
W & \hookrightarrow & B & \twoheadrightarrow & \nabla \\
\downarrow{\scriptstyle i} & & \| & & \downarrow \\
R & \hookrightarrow & B & \twoheadrightarrow & \overline{\nabla} \\
\downarrow & & & & \\
\text{cl}_S & & & &
\end{array} \tag{1.16}$$

We now choose $\mathbb{Q}G$-automorphisms α of $\mathbb{Q}W$ and β of $\mathbb{Q}R$ as well as $\mathbb{Q}G$-isomorphisms $\tilde{\alpha}$ and $\tilde{\beta}$ making the following diagrams commutative:

$$\begin{array}{ccccc}
\mathbb{Q}(E_S \oplus C) & \hookrightarrow & \mathbb{Q}(E_S \oplus C \oplus W) & \twoheadrightarrow & \mathbb{Q}W \\
\| & & \downarrow{\scriptstyle \tilde{\alpha}} & & \downarrow{\scriptstyle \alpha} \\
\mathbb{Q}(E_S \oplus C) & \hookrightarrow & \mathbb{Q}(A \oplus C) & \twoheadrightarrow & \mathbb{Q}W
\end{array} \tag{1.17}$$

$$\begin{array}{ccccc} \mathbb{Q}R & \hookrightarrow & \mathbb{Q}B & \twoheadrightarrow & \mathbb{Q}\overline{\nabla} \\ \downarrow \beta & & \downarrow \tilde{\beta} & & \| \\ \mathbb{Q}R & \hookrightarrow & \mathbb{Q}(R \oplus \overline{\nabla}) & \twoheadrightarrow & \mathbb{Q}\overline{\nabla} \end{array} \tag{1.18}$$

In diagram (1.17) C is a free $\mathbb{Z}G$-module as in Lemma 1.2.1. The lower sequence derives from adding C to the left part of the Tate-sequence. The upper sequence is the canonical one as well as the lower sequence in (1.18). The upper sequence in (1.18) is extracted from (1.16).
Given a $\mathbb{Q}G$-isomorphism $\phi : \mathbb{Q}\overline{\nabla} \to \mathbb{Q}(E_S \oplus C)$ as in Lemma 1.2.1 we define a $\mathbb{Q}G$-isomorphism $\tilde{\phi}$ to be the composite map

$$\tilde{\phi} : \mathbb{Q}B \xrightarrow{\tilde{\beta}} \mathbb{Q}(R \oplus \overline{\nabla}) \xrightarrow{\mathrm{id}_R \oplus \phi} \mathbb{Q}(R \oplus E_S \oplus C) \tag{1.19}$$

$$\xrightarrow{i^{-1} \oplus \mathrm{id}_{E_S \oplus C}} \mathbb{Q}(W \oplus E_S \oplus C) \xrightarrow{\tilde{\alpha}} \mathbb{Q}(A \oplus C).$$

We define

$$\boxed{\Omega_\phi := (B, \tilde{\phi}, A \oplus C) - \partial[\mathbb{Q}W, \alpha] - \partial[\mathbb{Q}R, \beta] \in K_0(\mathbb{Z}G, \mathbb{Q}).} \tag{1.20}$$

REMARK.

(1) One can choose the isomorphisms α and β to be the identity on $\mathbb{Q}W$ and $\mathbb{Q}R$, respectively. Sometimes, however, it may be useful to choose injections $W \rightarrowtail W$ and $R \rightarrowtail R$, since we can actually build $\mathbb{Z}G$-diagrams corresponding to those in (1.17) and (1.18) in this case. These injections automatically become isomorphisms after tensoring with $\mathbb{Q}$. This also shows the analogy to the construction in [GRW].

(2) If S is large in the sense that all ramified primes lie in S and $\mathrm{cl}_S = 1$, our construction yields the Ω_ϕ of [GRW] if we choose α and β to be $\mathbb{Z}G$-injections homotopic to 0. We will see in the next section that the definition of Ω_ϕ is independent of the choice of α, β, $\tilde{\alpha}$ and $\tilde{\beta}$.

(3) The natural homomorphism $K_0(\mathbb{Z}G, \mathbb{Q}) \to K_0(\mathbb{Z}G)$ sends Ω_ϕ to Chinburg's $\Omega_3(L/K)$ (cf. [Ch2], p. 357 or [We]).

1.3 Independence of choices

In the preceding section we have defined an element Ω_ϕ attached to the following data (D):

- a finite Galois extension L/K of number fields with Galois group G,
- a finite G-invariant set S of places of L which contains all the infinite primes,
- a $\mathbb{Q}G$-isomorphism $\phi : \mathbb{Q}\overline{\nabla} \to \mathbb{Q}(E_S \oplus C)$, where $\overline{\nabla}$ is the kernel of the sequence (1.13) and C is a free $\mathbb{Z}G$-module of rank $|S^*_{\mathrm{ram}} \setminus (S \cap S_{\mathrm{ram}})^*|$ as in Lemma 1.2.1.

We have made some choices during the construction, so the aim of this section will be to prove the following theorem.

Theorem 1.3.1 *The data (D) uniquely determine an element* $\Omega_\phi \in K_0(\mathbb{Z}G, \mathbb{Q})$.

We divide the proof into two lemmas.

Lemma 1.3.2 *The definition of* Ω_ϕ *is independent of the choices of* α, β, $\tilde{\alpha}$ *and* $\tilde{\beta}$.

PROOF. If we take other isomorphisms α', β', $\tilde{\alpha}'$, $\tilde{\beta}'$ and set $\tau = \alpha^{-1} \circ \alpha'$, $\sigma = \beta' \circ \beta^{-1}$ and accordingly $\tilde{\tau} = \tilde{\alpha}^{-1} \circ \tilde{\alpha}'$, $\tilde{\sigma} = \tilde{\beta}' \circ \tilde{\beta}^{-1}$, we get equalities

$$[\mathbb{Q}W, \alpha'] - [\mathbb{Q}W, \alpha] = [\mathbb{Q}W, \tau] = [\mathbb{Q}(W \oplus E_S \oplus C), \tilde{\tau}] \qquad (1.21)$$

and

$$[\mathbb{Q}R, \beta'] - [\mathbb{Q}R, \beta] = [\mathbb{Q}R, \sigma] = [\mathbb{Q}(R \oplus \overline{\nabla}), \tilde{\sigma}] \qquad (1.22)$$

in $K_1(\mathbb{Q}G)$ as follows from the commutative diagrams

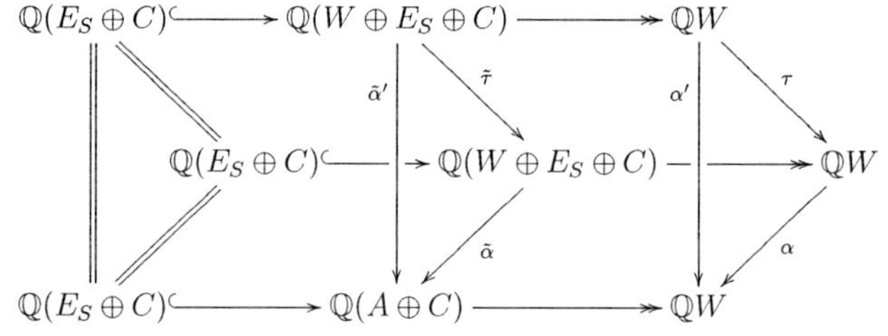

and

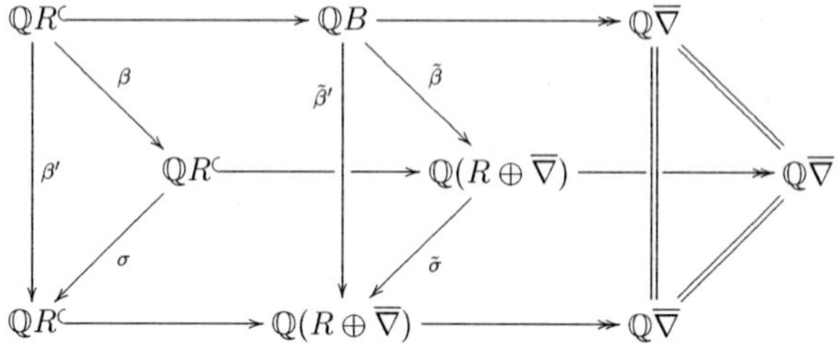

Let $\Psi = (B, \tilde{\phi}, A \oplus C)$ and $\Psi' = (B, \tilde{\phi}', A \oplus C)$, where $\tilde{\phi}$ arises from $\tilde{\alpha}$ and $\tilde{\beta}$, and $\tilde{\phi}'$ from $\tilde{\alpha}'$ and $\tilde{\beta}'$. We have to show that

$$\begin{aligned} \Psi' - \Psi &= \partial[\mathbb{Q}W, \alpha'] + \partial[\mathbb{Q}R, \beta'] - \partial[\mathbb{Q}W, \alpha] - \partial[\mathbb{Q}R, \beta] \\ &= \partial[\mathbb{Q}(W \oplus E_S \oplus C), \tilde{\tau}] + \partial[\mathbb{Q}(R \oplus \overline{\nabla}), \tilde{\sigma}]. \end{aligned}$$

by (1.21) and (1.22). For this, let

$$\gamma = (i^{-1} \oplus \mathrm{id}_{E_S \oplus C}) \circ (\mathrm{id}_R \oplus \phi) \circ \tilde{\beta} : \mathbb{Q}B \to \mathbb{Q}(W \oplus E_S \oplus C),$$

so $\tilde{\phi} = \tilde{\alpha} \circ \gamma$ and $\tilde{\phi}' = \tilde{\alpha}' \circ \gamma \circ \tilde{\beta}^{-1} \circ \tilde{\beta}'$ by (1.19). Now,

$$\begin{aligned} \Psi' - \Psi &= (B, \tilde{\phi}^{-1} \circ \tilde{\phi}', B) \\ &= \partial[\mathbb{Q}B, \tilde{\phi}^{-1} \circ \tilde{\phi}'] \\ &= \partial[\mathbb{Q}B, \gamma^{-1} \circ \tilde{\alpha}^{-1} \circ \tilde{\alpha}' \circ \gamma \circ \tilde{\beta}^{-1} \circ \tilde{\beta}'] \\ &= \partial[\mathbb{Q}B, \gamma^{-1} \circ \tilde{\tau} \circ \gamma] + \partial[\mathbb{Q}B, \tilde{\beta}^{-1} \circ \tilde{\sigma} \circ \tilde{\beta}] \\ &= \partial[\mathbb{Q}(W \oplus E_S \oplus C), \tilde{\tau}] + \partial[\mathbb{Q}(R \oplus \overline{\nabla}), \tilde{\sigma}], \end{aligned}$$

as desired. □

Secondly, we have to check:

Lemma 1.3.3 *The definition of Ω_ϕ is independent of the choice of the Tate-sequence.*

PROOF. It will be necessary to go through the details of the construction of Tate-sequences for small S (cf. [RW1]). Therefore, we review that construction and indicate all the choices made. Hereafter, we will discuss each of them separately.
Let S' be a finite set of places of L which contains $S \cup S_{\mathrm{ram}}$ and is large enough to generate the ideal class group of L, and such that $\bigcup_{\mathfrak{P} \in S'} G_{\mathfrak{P}} = G$ (1st choice). We fix a choice $*$ of a representative for each orbit of the action of G on the primes of L (2nd choice).
Let us denote the S-idèles of L by J_S, and the idèle class group of L by C_L. Choose an exact sequence

$$C_L \rightarrowtail \mathfrak{V} \twoheadrightarrow \Delta G$$

of $\mathbb{Z}G$-modules whose extension class maps to the global fundamental class $u_{L/K}$ via the isomorphism $\mathrm{Ext}^1_G(\Delta G, C_L) \simeq H^2(G, C_L)$. Locally, for each $\mathfrak{P} \in S'^*$ there are analogous exact sequences

$$L_{\mathfrak{P}}^{\times} \rightarrowtail V_{\mathfrak{P}} \twoheadrightarrow \Delta G_{\mathfrak{P}}$$

of $\mathbb{Z}G_{\mathfrak{P}}$-modules whose extension classes map to the local fundamental classes $u_{L_{\mathfrak{P}}/K_{\mathfrak{p}}}$ via the isomorphisms $\mathrm{Ext}^1_{G_{\mathfrak{P}}}(\Delta G_{\mathfrak{P}}, L_{\mathfrak{P}}^{\times}) \simeq H^2(G_{\mathfrak{P}}, L_{\mathfrak{P}}^{\times})$. We define $\mathbb{Z}G$-modules

$$\begin{array}{rcl} V_{S'} & = & \bigoplus_{\mathfrak{P}\in S'^*} \mathrm{ind}\,_{G_{\mathfrak{P}}}^{G} V_{\mathfrak{P}} \times \prod_{\mathfrak{P}\notin S'} U_{\mathfrak{P}} \\ W_{S'} & = & \bigoplus_{\mathfrak{P}\in S^*} \mathrm{ind}\,_{G_{\mathfrak{P}}}^{G} \Delta G_{\mathfrak{P}} \oplus \bigoplus_{\mathfrak{P}\in S'^*\setminus S^*} \mathrm{ind}\,_{G_{\mathfrak{P}}}^{G} W_{\mathfrak{P}}, \end{array} \tag{1.23}$$

where $U_{\mathfrak{P}}$ are the units of $L_{\mathfrak{P}}$, and $W_{\mathfrak{P}}$ is the inertial lattice of the extension $L_{\mathfrak{P}}/K_{\mathfrak{p}}$ (see (1.11)). Starting with the local sequence above, the pushout along the normalized valuation $v_{\mathfrak{P}} : L_{\mathfrak{P}}^{\times} \twoheadrightarrow \mathbb{Z}$ yields the commutative diagram (cf. [We], p. 42):

$$\begin{array}{ccccc} U_{\mathfrak{P}} & = & U_{\mathfrak{P}} & & \\ \downarrow & & \downarrow & & \\ L_{\mathfrak{P}}^{\times} & \hookrightarrow & V_{\mathfrak{P}} & \twoheadrightarrow & \Delta G_{\mathfrak{P}} \\ \downarrow{\scriptstyle v_{\mathfrak{P}}} & & \downarrow & & \| \\ \mathbb{Z} & \hookrightarrow & W_{\mathfrak{P}} & \twoheadrightarrow & \Delta G_{\mathfrak{P}} \end{array} \tag{1.24}$$

Thus, we locally get exact sequences $U_{\mathfrak{P}} \rightarrowtail V_{\mathfrak{P}} \twoheadrightarrow W_{\mathfrak{P}}$, and hence an exact sequence

$$J_S \rightarrowtail V_{S'} \twoheadrightarrow W_{S'} \tag{1.25}$$

of $\mathbb{Z}G$-modules. By Theorem 1 in [RW1] we find a surjective $\mathbb{Z}G$-homomorphism θ (3rd choice) which fits into the diagram

$$\begin{array}{ccccc} J_S & \hookrightarrow & V_{S'} & \twoheadrightarrow & W_{S'} \\ \downarrow & & \downarrow{\scriptstyle \theta} & & \downarrow{\scriptstyle c} \\ C_L & \hookrightarrow & \mathfrak{V} & \twoheadrightarrow & \Delta G \end{array} \tag{1.26}$$

where c is induced by the inclusions $\Delta G_{\mathfrak{P}} \subset \Delta G$ for $\mathfrak{P} \in S^*$ and by

$$W_{\mathfrak{P}} \twoheadrightarrow \Delta G_{\mathfrak{P}} \subset \Delta G$$

for $\mathfrak{P} \in S'^* \setminus S^*$.

There are no further choices made in the construction; nevertheless, we continue with its description for later use.

Since the left vertical map $J_S \to C_L$ has kernel E_S and cokernel cl_S, the S-class group of L, the Snake Lemma produces an exact sequence

$$E_S \rightarrowtail A_\theta \to R_{S'} \twoheadrightarrow \mathrm{cl}_S \tag{1.27}$$

of $\mathbb{Z}G$-modules, where A_θ is c.t. and $R_{S'}$ is a $\mathbb{Z}G$-lattice. Now we combine various diagrams for three types of primes $\mathfrak{P} \in S'^*$ (see [RW1], p. 157 or Proposition 1.5.4 for the first, the others are clear).
Type 1: $\mathfrak{P} \in S^*_{\mathrm{ram}} \setminus (S \cap S_{\mathrm{ram}})^*$

$$\begin{array}{ccccc} W_{\mathfrak{P}} & \hookrightarrow & \mathbb{Z}G^2_{\mathfrak{P}} & \twoheadrightarrow & W^0_{\mathfrak{P}} \\ \downarrow & & \downarrow & & \downarrow \\ \Delta G & \hookrightarrow & \mathbb{Z}G & \twoheadrightarrow & \mathbb{Z} \end{array} \tag{1.28}$$

Type 2: $\mathfrak{P} \in S^*$

$$\begin{array}{ccccc} \Delta G_{\mathfrak{P}} & \hookrightarrow & \mathbb{Z}G_{\mathfrak{P}} & \twoheadrightarrow & \mathbb{Z} \\ \downarrow & & \downarrow & & \downarrow \\ \Delta G & \hookrightarrow & \mathbb{Z}G & \twoheadrightarrow & \mathbb{Z} \end{array} \tag{1.29}$$

Type 3: $\mathfrak{P} \in S'^* \setminus (S^* \cup S^*_{\mathrm{ram}})$

$$\begin{array}{ccccc} W_{\mathfrak{P}} & \xrightarrow{\simeq} & \mathbb{Z}G_{\mathfrak{P}} & \longrightarrow & 0 \\ \downarrow{\scriptstyle 0} & & \downarrow{\scriptstyle 0} & & \downarrow \\ \Delta G & \hookrightarrow & \mathbb{Z}G & \twoheadrightarrow & \mathbb{Z} \end{array} \tag{1.30}$$

If we define

$$N_{S'} = \bigoplus_{\mathfrak{P} \text{ of type 1}} \mathrm{ind}\,^G_{G_{\mathfrak{P}}}(\mathbb{Z}G^2_{\mathfrak{P}}) \oplus \bigoplus_{\mathfrak{P} \text{ of type 2 or 3}} \mathrm{ind}\,^G_{G_{\mathfrak{P}}} \mathbb{Z}G_{\mathfrak{P}},$$

$$M^* = \bigoplus_{\mathfrak{P} \text{ of type 1}} \mathrm{ind}\,^G_{G_{\mathfrak{P}}} W^0_{\mathfrak{P}} \oplus \bigoplus_{\mathfrak{P} \text{ of type 2}} \mathrm{ind}\,^G_{G_{\mathfrak{P}}} \mathbb{Z},$$

the three diagrams above yield

$$\begin{array}{ccccc} R_{S'} & \hookrightarrow & B_{S'} & \twoheadrightarrow & \nabla^* \\ \downarrow & & \downarrow & & \downarrow \\ W_{S'} & \hookrightarrow & N_{S'} & \twoheadrightarrow & M^* \\ \downarrow & & \downarrow & & \downarrow \\ \Delta G & \hookrightarrow & \mathbb{Z}G & \twoheadrightarrow & \mathbb{Z} \end{array} \tag{1.31}$$

Observe that $B_{S'}$ is projective, since $N_{S'}$ is $\mathbb{Z}G$-free. As a last step we take the pushout of the upper sequence in (1.31) along the surjection $R_{S'} \twoheadrightarrow \mathrm{cl}_S$ in (1.27):

$$\begin{array}{ccccc} R_{S'} & \hookrightarrow & B_{S'} & \twoheadrightarrow & \overline{\nabla}^* \\ \downarrow & & \downarrow & & \| \\ \mathrm{cl}_S & \hookrightarrow & \nabla_\theta & \twoheadrightarrow & \overline{\nabla}^* \end{array} \tag{1.32}$$

Together with (1.27) this yields a Tate-sequence for S:

$$E_S \rightarrowtail A_\theta \to B_{S'} \twoheadrightarrow \nabla_\theta$$

Before we go into the discussion of choices, we insert the following proposition, which will be useful in the following.

Proposition 1.3.4 *Underlying the data (D), assume that there are two Tate-sequences for S as shown in the diagram:*

$$\begin{array}{ccccccc} E_S & \hookrightarrow & A & \to & B & \twoheadrightarrow & \nabla \\ \| & & \downarrow a & & \downarrow b & & \downarrow h \, \simeq \\ E_S & \hookrightarrow & A' & \to & B' & \twoheadrightarrow & \nabla' \\ & & \downarrow & & \downarrow & & \\ & & P & = & P & & \end{array}$$

Suppose that P is $\mathbb{Z}G$-projective and the isomorphism h fits into a diagram

$$\begin{array}{ccccc} \mathrm{cl}_S & \hookrightarrow & \nabla & \twoheadrightarrow & \overline{\nabla} \\ \| & & \downarrow h \, \simeq & & \downarrow \overline{h} \, \simeq \\ \mathrm{cl}_S & \hookrightarrow & \nabla' & \twoheadrightarrow & \overline{\nabla} \end{array} \tag{1.33}$$

Then we have an equality

$$\Omega_\phi = \Omega'_{\phi\overline{h}^{-1}},$$

where Ω_ϕ and $\Omega'_{\phi\overline{h}^{-1}}$ arise from the upper and the lower Tate-sequence, respectively.
In particular, if $\overline{h} = \mathrm{id}_{\overline{\nabla}}$, we have $\Omega_\phi = \Omega'_\phi$.

REMARK. In [RW1] an isomorphism h as in diagram (1.33) satisfying $\overline{h} = \mathrm{id}_{\overline{\nabla}}$ is called admissible (cf. Theorem 4 in loc. cit.).

PROOF. Since P is projective, we have compatible isomorphisms $A' \simeq A\oplus P$ and $B' \simeq B \oplus P$. Replace the upper Tate-sequence by

$$E_S \rightarrowtail A\oplus P \to B\oplus P \twoheadrightarrow \nabla.$$

This clearly leaves Ω_ϕ unchanged, since we may replace the isomorphisms $\alpha,\ \tilde{\alpha},\ \beta,\ \tilde{\beta}$ by $\alpha\oplus\mathrm{id}_P,\ \tilde{\alpha}\oplus\mathrm{id}_P,\ \beta\oplus\mathrm{id}_P,\ \tilde{\beta}\oplus\mathrm{id}_P$. Hence, we may assume $P=0$.
We get a commutative diagram, in which all modules are invisibly tensored with $\mathbb{Q}$:

$$\begin{array}{ccccccccc}
B & \multicolumn{7}{c}{\xrightarrow{\qquad\qquad\qquad\qquad\tilde{\phi}\qquad\qquad\qquad\qquad}} & A\oplus C \\
B & \longrightarrow & R\oplus\overline{\nabla} & \xrightarrow{\mathrm{id}_R\oplus\phi} & R\oplus E_S\oplus C & \longrightarrow & W\oplus E_S\oplus C & \longrightarrow & A\oplus C \\
\Big\downarrow{\scriptstyle b} & & \Big\downarrow{\scriptstyle r\oplus\overline{h}} & & \Big\downarrow{\scriptstyle r\oplus\mathrm{id}_{E_S\oplus C}} & & \Big\downarrow{\scriptstyle w\oplus\mathrm{id}_{E_S\oplus C}} & & \Big\downarrow{\scriptstyle a\oplus\mathrm{id}_C} \\
B' & \longrightarrow & R'\oplus\overline{\nabla} & \xrightarrow{\mathrm{id}_{R'}\oplus\phi\overline{h}^{-1}} & R'\oplus E_S\oplus C & \longrightarrow & W'\oplus E_S\oplus C & \longrightarrow & A'\oplus C \\
B' & \multicolumn{7}{c}{\xrightarrow[\qquad\qquad\qquad\qquad\widetilde{\phi\overline{h}^{-1}}\qquad\qquad\qquad\qquad]{}} & A'\oplus C
\end{array}$$

The $\mathbb{Z}G$-lattices R, W and R', W' are those of diagram (1.16) for the upper and lower Tate-sequence, respectively. The isomorphisms r and w are induced by b and h. Since the isomorphisms b and $a\oplus\mathrm{id}_C$ already exist at $\mathbb{Z}G$-level, we are done. □

For the proof of Lemma 1.3.3 we have to go through the proofs in [RW1].

Assertion 1.3.5 *Ω_ϕ is independent of the choice of the surjection θ.*

Assume that we have taken another surjection θ'. We indicate the modules involved by subscripts θ resp. θ' if they occur in the construction via θ resp. θ'. In [RW1], p. 171 it is shown that there is a commutative diagram

$$\begin{array}{ccccccc}
E_S & \hookrightarrow & A^+_{\theta'} & \longrightarrow & B'^+_{S'} & \twoheadrightarrow & \nabla^+_{\theta'} \\
\Big\| & & \Big\downarrow{\scriptstyle\simeq} & & \Big\downarrow{\scriptstyle\simeq} & & \Big\downarrow{\scriptstyle\simeq} \\
E_S & \hookrightarrow & A^+_{\theta} & \longrightarrow & B^+_{S'} & \twoheadrightarrow & \nabla^+_{\theta}
\end{array}$$

where the rows are Tate-sequences, and we have adopted the local notation. ∇^+_θ is isomorphic to ∇_θ via an admissible isomorphism, and the difference between the corresponding Tate-sequences is described via a commutative diagram as in Proposition 1.3.4. Since the same is true for $\nabla^+_{\theta'}$ and $\nabla_{\theta'}$, Proposition 1.3.4 implies Assertion 1.3.5.

Assertion 1.3.6 *Ω_ϕ is independent of the choice of S'.*

Let S'' be another set of places of L which satisfies the conditions as described at the beginning of Lemma 1.3.3. We may assume that $S' \subset S''$. Hence, there is an exact sequence of $\mathbb{Z}G$-modules

$$W_{S'} \rightarrowtail W_{S''} \twoheadrightarrow P = \bigoplus_{\mathfrak{P} \in S''^* \setminus S'^*} \operatorname{ind}_{G_{\mathfrak{P}}}^{G} W_{\mathfrak{P}},$$

where P is $\mathbb{Z}G$-free. As one learns from [RW1], p. 174, this gives rise to a diagram as in Proposition 1.3.4 with an admissible isomorphism.

We are left we the dependence on the choice of $*$. Let $\diamond$ be a second choice of G-orbit representatives of primes of L. For each $\mathfrak{P}$ distinguished by $*$ let $x_{\mathfrak{P}} \in G$ have the property that $x_{\mathfrak{P}}\mathfrak{P} = \mathfrak{P}'$ is distinguished by $\diamond$. As described in [RW1] such a system X of elements of G induces a transport $* \to \diamond$ and natural $\mathbb{Z}G$-module transport maps

$$X : W^* \to W^\diamond,\ \overline{\nabla}^* \to \overline{\nabla}^\diamond.$$

Hence, an isomorphism $\phi_* : \mathbb{Q}\overline{\nabla}^* \to \mathbb{Q}(E_S \oplus C)$ induces an isomorphism

$$\phi_\diamond = \phi_* \circ X^{-1} : \mathbb{Q}\overline{\nabla}^\diamond \to \mathbb{Q}(E_S \oplus C).$$

Assertion 1.3.7 *With the above notation we have:* $\Omega_{\phi_*} = \Omega_{\phi_\diamond}$.

As shown in [RW1], p. 176 et seq. one has a commutative diagram

$$\begin{array}{ccccccc}
E_S & \hookrightarrow & A^* & \longrightarrow & B^* & \twoheadrightarrow & \nabla^* \\
\| & & \downarrow \simeq & & \downarrow \simeq & & h \downarrow \simeq \\
E_S & \hookrightarrow & A^\diamond & \longrightarrow & B^\diamond & \longrightarrow & \nabla^\diamond
\end{array}$$

where the isomorphism h is X-admissible, i.e. it fits into a diagram

$$\begin{array}{ccccc}
\mathrm{cl}_S & \hookrightarrow & \nabla^* & \twoheadrightarrow & \overline{\nabla}^* \\
\| & & \downarrow h & & \downarrow X \\
\mathrm{cl}_S & \hookrightarrow & \nabla^\diamond & \twoheadrightarrow & \overline{\nabla}^\diamond
\end{array}$$

Hence, we have $\mathbb{Z}G$-isomorphisms $B^* \simeq B^\diamond$ and $A^* \oplus C \simeq A^\diamond \oplus C$, which commute with $\tilde{\phi}_*$ and $\tilde{\phi}_\diamond$ after tensoring with $\mathbb{Q}$:

$$\begin{array}{ccc}
\mathbb{Q}B^* & \xrightarrow{\tilde{\phi}_*} & \mathbb{Q}(A^* \oplus C) \\
\downarrow \simeq & & \downarrow \simeq \\
\mathbb{Q}B^\diamond & \xrightarrow{\tilde{\phi}_\diamond} & \mathbb{Q}(A^\diamond \oplus C)
\end{array}$$

Thus, $\Omega_{\phi_*} = \Omega_{\phi_\diamond}$ by Lemma 1.1.6. □

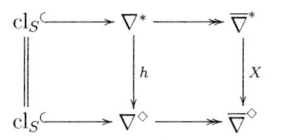

1.4 Basic properties of Ω_ϕ

In this section we discuss variance of the isomorphism ϕ and of the set of places S. The most interesting (and most complicated) case is, how Ω_ϕ varies if S is enlarged by ramified primes. Before going into this, however, we give an alternative definition of Ω_ϕ.

Keeping the notation of the preceding section we start with a $\mathbb{Q}G$-isomorphism $\phi' : \mathbb{Q}\nabla \to \mathbb{Q}(E_S \oplus C)$, which exists due to the exact sequence

$$\mathrm{cl}_S \hookrightarrow \nabla \xrightarrow{\pi_\nabla} \overline{\nabla}$$

and Lemma 1.2.1. We choose $\mathbb{Q}G$-automorphisms α and β_W of $\mathbb{Q}W$, where W is the $\mathbb{Z}G$-lattice defined via splitting the Tate-sequence into two parts (cf. (1.15)). Choose $\tilde{\alpha}$ as in (1.17) and a $\mathbb{Q}G$-isomorphism $\tilde{\beta}_W$ such that the following diagram commutes:

$$\begin{array}{ccccc} \mathbb{Q}W & \hookrightarrow & \mathbb{Q}B & \twoheadrightarrow & \mathbb{Q}\nabla \\ \downarrow \beta_W & & \downarrow \tilde{\beta}_W & & \| \\ \mathbb{Q}W & \hookrightarrow & \mathbb{Q}(W \oplus \nabla) & \twoheadrightarrow & \mathbb{Q}\nabla \end{array}$$

We now define the $\mathbb{Q}G$-isomorphism $\tilde{\phi}' : \mathbb{Q}B \to \mathbb{Q}(A \oplus C)$ to be the composite map

$$\tilde{\phi}' : \mathbb{Q}B \xrightarrow{\tilde{\beta}_W} \mathbb{Q}(W \oplus \nabla) \tag{1.34}$$

$$\xrightarrow{\mathrm{id}_W \oplus \phi'} \mathbb{Q}(W \oplus E_S \oplus C) \xrightarrow{\tilde{\alpha}} \mathbb{Q}(A \oplus C)$$

Finally, we define

$$\hat{\Omega}_{\phi'} := (B, \tilde{\phi}', A \oplus C) - \partial[\mathbb{Q}W, \alpha \circ \beta_W].$$

Proposition 1.4.1 *Assume that we have given a set of data (D), where $\phi = \phi' \circ \pi_\nabla^{-1}$ for a $\mathbb{Q}G$-isomorphism $\phi' : \mathbb{Q}\nabla \to \mathbb{Q}(E_S \oplus C)$. Then we have an equality*

$$\Omega_\phi = \hat{\Omega}_{\phi'}.$$

PROOF. We have the following commutative diagram:

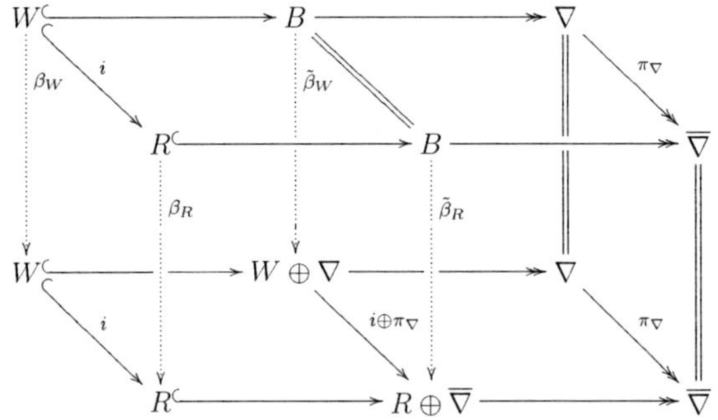

Here, $\beta_R = i \circ \beta_W \circ i^{-1}$ and $\tilde\beta_R = (i \oplus \pi_\nabla) \circ \tilde\beta_W$. The dotted arrows only exist after tensoring with $\mathbb{Q}$; the top face is the main part of diagram (1.16). All vertical maps as well as i and π_∇ become isomorphisms after tensoring with $\mathbb{Q}$. Hence, we have $[\mathbb{Q}R, \beta_R] = [\mathbb{Q}W, \beta_W]$ in $K_1(\mathbb{Q}G)$. Moreover, we have a commutative diagram, in which all occurring modules are invisibly tensored with $\mathbb{Q}$:

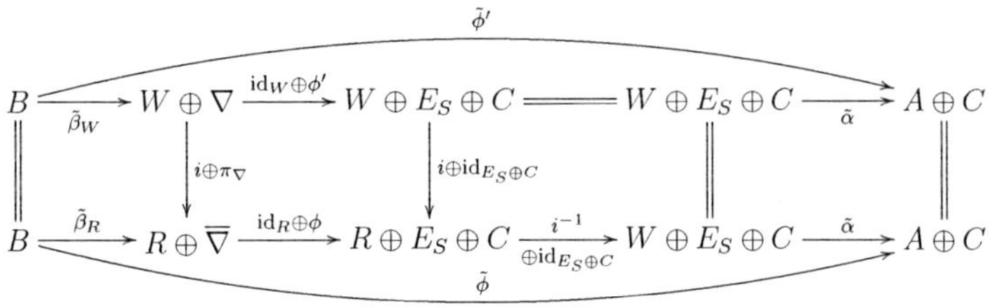

Therefore

$$\begin{aligned} \Omega_\phi &= (B, \tilde\phi, A \oplus C) - \partial[\mathbb{Q}W, \alpha] - \partial[\mathbb{Q}R, \beta_R] \\ &= (B, \tilde\phi', A \oplus C) - \partial[\mathbb{Q}W, \alpha] - \partial[\mathbb{Q}W, \beta_W] \\ &= \hat\Omega_{\phi'}. \end{aligned}$$

This proves the proposition. □

REMARK.

(1) The above definition has the advantage that one does not need the $\mathbb{Z}G$-lattice R, but the disadvantage that one cannot work at $\mathbb{Z}G$-level: In general there do not exist injections $\nabla \rightarrowtail E_S \oplus C$. By contrast, we can always find injections $\overline{\nabla} \rightarrowtail E_S \oplus C$, since $\overline{\nabla}$ has no $\mathbb{Z}$-torsion.

(2) Proposition 1.4.1 shows that we can describe Ω_ϕ via refined Euler characteristics (but we will make no use of this fact): Consider the perfect complex

$$C^\cdot : \ldots \to 0 \to A \to B \to 0 \to \ldots$$

where the position of A is in degree zero and the map $A \to B$ is taken from the Tate sequence. Then $\Omega_\phi = \hat{\Omega}_{\phi'} = \chi_{\mathbb{Z}G,\mathbb{Q}G}(C^\cdot, (\phi')^{-1})$.

The following proposition describes variance with ϕ and is the analogue to Proposition 1 in [GRW].

Proposition 1.4.2 *Fix a set of data (D), and let $\phi' : \mathbb{Q}\overline{\nabla} \to \mathbb{Q}(E_S \oplus C)$ be another $\mathbb{Q}G$-isomorphism. Then*

$$\Omega_{\phi'} - \Omega_\phi = \partial[\mathbb{Q}\overline{\nabla}, \phi^{-1} \circ \phi'].$$

In particular, $\Omega_{\phi'} - \Omega_\phi$ has representing homomorphism

$$\chi \mapsto \det(\phi^{-1} \circ \phi' | \mathrm{Hom}_{\mathbb{C}G}(V_\chi, \mathbb{C}\overline{\nabla})),$$

where V_χ is a $\mathbb{C}G$-module with character χ.

PROOF. If we build $\tilde{\phi}$ and $\tilde{\phi}'$ using the same maps α, β, $\tilde{\alpha}$, $\tilde{\beta}$, there is a commutative diagram

$$\begin{array}{ccccccccc}
B & \xrightarrow{\tilde{\beta}} & R \oplus \overline{\nabla} & \xrightarrow{\mathrm{id}_R \oplus \phi} & R \oplus E_S \oplus C & \longrightarrow & W \oplus E_S \oplus C & \longrightarrow & A \oplus C \\
\uparrow {\scriptstyle \tilde{\beta}^{-1}\circ\gamma\circ\tilde{\beta}} & & \uparrow {\scriptstyle \gamma} & & \| & & \| & & \| \\
B & \xrightarrow{\tilde{\beta}} & R \oplus \overline{\nabla} & \xrightarrow{\mathrm{id}_R \oplus \phi'} & R \oplus E_S \oplus C & \longrightarrow & W \oplus E_S \oplus C & \longrightarrow & A \oplus C
\end{array}$$

(with the top row composing to $\tilde{\phi}$ and the bottom row composing to $\tilde{\phi}'$)

where $\gamma = \mathrm{id}_R \oplus (\phi^{-1} \circ \phi')$ and all modules are invisibly tensored with $\mathbb{Q}$. Now,

$$\begin{aligned}
\Omega_{\phi'} - \Omega_\phi &= (B, \tilde{\phi}', A \oplus C) - (B, \tilde{\phi}, A \oplus C) \\
&= (B, \tilde{\phi}^{-1} \circ \tilde{\phi}', B) \\
&= \partial[\mathbb{Q}B, \tilde{\phi}^{-1} \circ \tilde{\phi}'] \\
&= \partial[\mathbb{Q}B, \tilde{\beta}^{-1} \circ \gamma \circ \tilde{\beta}] \\
&= \partial[\mathbb{Q}(R \oplus \overline{\nabla}), \gamma] \\
&= \partial[\mathbb{Q}\overline{\nabla}, \phi^{-1} \circ \phi'],
\end{aligned}$$

as desired. □

Our next task is to enlarge S by a ramified prime $\mathfrak{P}_0$, i.e. $\mathfrak{P}_0 \in S_{\mathrm{ram}}$, but $\mathfrak{P}_0 \notin S$. We may assume $\mathfrak{P}_0 \in S^*_{\mathrm{ram}}$.

Note that some of the ideas in what follows are taken from [Gr3], where the author assumes the validity of the LRNC for an abelian CM-extension L/K to compute the Fitting ideal of $(\mathrm{cl}_L^-)^\vee$, the Pontryagin dual of the minus class group of L. For this, he connects a Tate-sequence for a large set S of places of L to a Tate-sequence for S_∞. In what follows here, some of the maps between Tate-sequences are inspired by the corresponding maps in [Gr3]. But some of the diagrams in loc. cit. only commute on minus parts owing to the purpose of this paper; so we have to modify the construction in order to achieve commutative diagrams in general. Moreover, the author does not introduce an element like Ω_ϕ, nor he gives a definition of a modified Stark-Tate regulator, as we intend to do in the next section. Indeed, it considerably simplifies matters if one restricts to minus parts, since the infinite primes pleasantly drop out.

We set $S_0 := S \cup G\mathfrak{P}_0$ and we intend to indicate each module by a subscript S resp. S_0 (or simply a subscript 0) if it is not clear to which (construction of a) Tate-sequence it belongs.

The dual of the sequence (1.12) for the prime $\mathfrak{P}_0$, namely

$$\Delta G^0_{\mathfrak{P}_0} \rightarrowtail W^0_{\mathfrak{P}_0} \twoheadrightarrow \mathbb{Z}^0 = \mathbb{Z},$$

yields the following commutative diagram:

$$\begin{array}{ccccc}
\mathrm{ind}\,^G_{G_{\mathfrak{P}_0}} \Delta G^0_{\mathfrak{P}_0} & = & \mathrm{ind}\,^G_{G_{\mathfrak{P}_0}} \Delta G^0_{\mathfrak{P}_0} & & \\
\downarrow & & \downarrow & & \\
\overline{\nabla}_S & \hookrightarrow & \mathbb{Z}S \oplus \bigoplus_{\mathfrak{p}\in S^*_{\mathrm{ram}}\setminus(S\cap S_{\mathrm{ram}})^*} \mathrm{ind}\,^G_{G_\mathfrak{P}} W^0_\mathfrak{P} & \twoheadrightarrow & \mathbb{Z} \\
\downarrow & & \downarrow & & \| \\
\overline{\nabla}_{S_0} & \hookrightarrow & \mathbb{Z}S_0 \oplus \bigoplus_{\mathfrak{p}\in S^*_{\mathrm{ram}}\setminus(S_0\cap S_{\mathrm{ram}})^*} \mathrm{ind}\,^G_{G_\mathfrak{P}} W^0_\mathfrak{P} & \twoheadrightarrow & \mathbb{Z}
\end{array}$$

We extract the left column and use (1.1) to get an exact sequence

$$\mathbb{Z}G/N_{G_{\mathfrak{P}_0}} \hookrightarrow \overline{\nabla}_S \xrightarrow{\pi_{\overline{\nabla}}} \overline{\nabla}_{S_0}. \tag{1.35}$$

Let $h_L = |\mathrm{cl}_L|$ be the class number of L and choose a positive integer h such that $h_L | h$. Then $\mathfrak{P}_0^h$ is principal, and we find an S_0-unit $u_{\mathfrak{P}_0}$ which satisfies $v_{\mathfrak{P}_0}(u_{\mathfrak{P}_0}) = h$ and $v_\mathfrak{P}(u_{\mathfrak{P}_0}) = 0$ for all non-archimedean primes $\mathfrak{P} \neq \mathfrak{P}_0$. Here, $v_\mathfrak{P}$ denotes the normalized valuation at $\mathfrak{P}$. Let us define a map (which is the map β in [Gr3])

$$u_0 : \mathbb{Z}G \to E_{S_0},\ 1 \mapsto u_{\mathfrak{P}_0}.$$

Then we have a left exact sequence

$$\Delta G_{\mathfrak{P}_0} \cdot \mathbb{Z}G \xrightarrow{(-u_0,\mathrm{id})} E_S \oplus \mathbb{Z}G \xrightarrow{(\mathrm{id},u_0)} E_{S_0}, \tag{1.36}$$

since $u_{\mathfrak{P}_0}^x \in E_S$ if and only if $x \equiv 0 \bmod G_{\mathfrak{P}_0}$. We have a $\mathbb{Q}G$-isomorphism

$$\begin{array}{rrcl} \phi' : & \mathbb{Q}G/N_{G_{\mathfrak{P}_0}} & \to & \Delta G_{\mathfrak{P}_0} \cdot \mathbb{Q}G, \\ & 1 \bmod N_{G_{\mathfrak{P}_0}} & \mapsto & 1 - \frac{1}{|G_{\mathfrak{P}_0}|} N_{G_{\mathfrak{P}_0}}. \end{array} \tag{1.37}$$

Let C_0 be a free $\mathbb{Z}G$-module of rank $|S_{\mathrm{ram}}^* \setminus (S_0 \cap S_{\mathrm{ram}})^*|$, and start with a $\mathbb{Q}G$-isomorphism $\phi_0 : \mathbb{Q}\overline{\nabla}_{S_0} \to \mathbb{Q}(E_{S_0} \oplus C_0)$. Then one can always find a $\mathbb{Q}G$-isomorphism ϕ fitting in a commutative diagram

$$\begin{array}{ccc} \mathbb{Q}G/N_{G_{\mathfrak{P}_0}} & \xrightarrow{\phi'} & \Delta G_{\mathfrak{P}_0} \cdot \mathbb{Q}G \\ \downarrow & & \downarrow {\scriptstyle (-u_0,\mathrm{id},0)} \\ \mathbb{Q}\overline{\nabla}_S & \xrightarrow{\phi} & \mathbb{Q}(E_S \oplus \mathbb{Z}G \oplus C_0) \\ \downarrow & & \downarrow {\scriptstyle (\mathrm{id},u_0,\mathrm{id}_{C_0})} \\ \mathbb{Q}\overline{\nabla}_{S_0} & \xrightarrow{\phi_0} & \mathbb{Q}(E_{S_0} \oplus C_0) \end{array} \tag{1.38}$$

Here, the two columns derive from the sequences (1.35) and (1.36), since the second map in (1.36) has finite cokernel.

We are ready to prove

Theorem 1.4.3 *Fix a set of data (D). Let $\mathfrak{P}_0$ be a prime not in S which ramifies in L/K and h an integral multiple of h_L, the class number of L. Assume that there is a $\mathbb{Q}G$-isomorphism ϕ_0 that fits into diagram (1.38). Then we have an equality*

$$\Omega_{\phi_0} - \Omega_\phi = \partial[\mathrm{ind}_{G_{\mathfrak{P}_0}}^G \mathbb{Q}, -h|G_{\mathfrak{P}_0}|].$$

In particular, $\Omega_{\phi_0} - \Omega_\phi$ has representing homomorphism

$$\chi \mapsto (-h|G_{\mathfrak{P}_0}|)^{\dim V_\chi^{G_{\mathfrak{P}_0}}},$$

where V_χ is a $\mathbb{C}G$-module with character χ.

PROOF. It is unavoidable to go through the whole construction of Tate sequences for small sets of places. We expand the notation of the proof of Lemma (1.3.3).
For this, let S' be a finite set of places of L which contains $S_0 \cup S_{\mathrm{ram}}$ and is large enough to generate the ideal class group of L, and such that $\bigcup_{\mathfrak{P} \in S'} G_{\mathfrak{P}} = G$. According to the definition of $W_{S'}$ let

$$W_{S',0} = \bigoplus_{\mathfrak{P} \in S_0^*} \mathrm{ind}_{G_{\mathfrak{P}}}^G \Delta G_{\mathfrak{P}} \oplus \bigoplus_{\mathfrak{P} \in S'^* \setminus S_0^*} \mathrm{ind}_{G_{\mathfrak{P}}}^G W_{\mathfrak{P}}.$$

Due to (1.12) we have an exact sequence

$$\operatorname{ind}_{G_{\mathfrak{P}_0}}^{G} \mathbb{Z} \hookrightarrow W_{S'} \twoheadrightarrow W_{S',0}.$$

The first step in the construction now yields a commutative diagram:

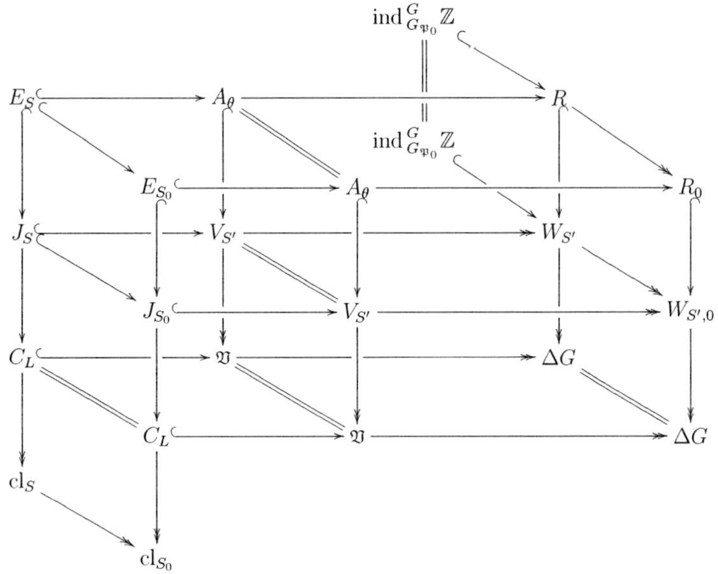

(1.39)

Due to the Snake Lemma we can extract from this the following diagram, where we split the two four-term sequences into short exact sequences:

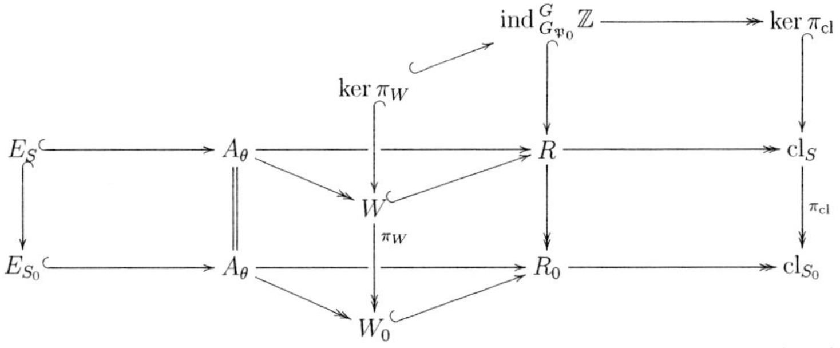

(1.40)

In analogy to the modules M and $N_{S'}$ we define

$$M_0 = \bigoplus_{\mathfrak{P}\in S_0^*} \operatorname{ind}_{G_{\mathfrak{P}}}^{G} \mathbb{Z} \oplus \bigoplus_{\mathfrak{P}\in S_{\mathrm{ram}}^*\setminus(S_0\cap S_{\mathrm{ram}})^*} \operatorname{ind}_{G_{\mathfrak{P}}}^{G} W_{\mathfrak{P}}^0,$$

$$N_{S',0} = \bigoplus_{\mathfrak{P}\in(S'\setminus S_{\mathrm{ram}})^*\cup S_0^*} \operatorname{ind}_{G_{\mathfrak{P}}}^{G} \mathbb{Z}G_{\mathfrak{P}} \oplus \bigoplus_{\mathfrak{P}\in S_{\mathrm{ram}}^*\setminus(S_0\cap S_{\mathrm{ram}})^*} \operatorname{ind}_{G_{\mathfrak{P}}}^{G} (\mathbb{Z}G_{\mathfrak{P}}^2)$$

and the second step in the construction yields a commutative diagram

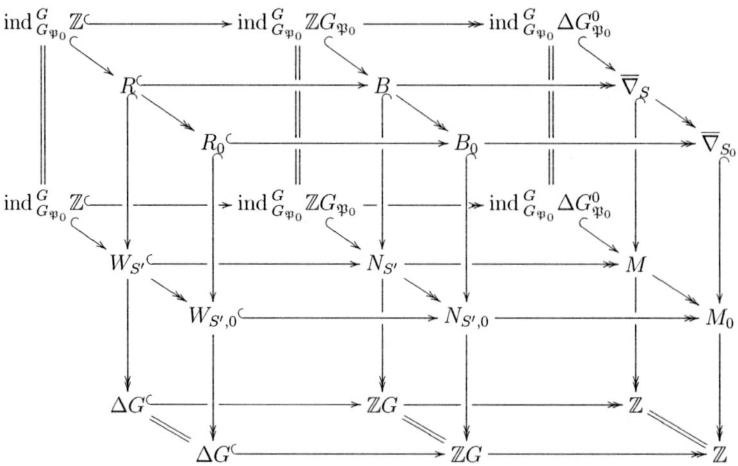

(1.41)

We choose the endomorphism β in diagram (1.18) and the endomorphism β_0 corresponding to R_0 to be the identity. We get the following commutative diagram in which we have invisibly tensored with $\mathbb{Q}$, and whose roof is the same as in the diagram above:

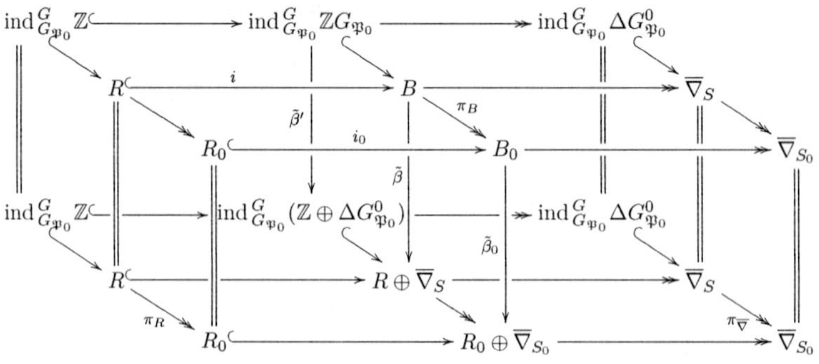

Note that we have labelled some of the maps in the above diagram.

We choose the isomorphisms $\tilde{\beta}$ and $\tilde{\beta}_0$ such that the projection $R \oplus \overline{\nabla}_S \twoheadrightarrow R_0 \oplus \overline{\nabla}_{S_0}$ is given by $\pi_R \oplus \pi_{\overline{\nabla}}$. This is possible by Lemma 1.1.2, since we may define these isomorphisms via commutative sections $\sigma : B \to R$ and $\sigma_0 : B_0 \to R_0$ of the injections i and i_0, respectively.

We also choose the automorphisms α of $\mathbb{Q}W$ and α_0 of $\mathbb{Q}W_0$ to be the identity. Let us abbreviate the map $(\mathrm{id}, u_0, \mathrm{id}_{C_0}) : E_S \oplus \mathbb{Z}G \oplus C_0 \to E_{S_0} \oplus C_0$ by δ and set $C := \mathbb{Z}G \oplus C_0$. Furthermore, let us write ι for the inclusion $E_{S_0} \rightarrowtail A := A_\theta$ and define $\pi_A := (\mathrm{id}_A + \iota u_0, \mathrm{id}_{C_0}) : A \oplus C \to A \oplus C_0$. Then we have a commutative diagram, where we have once more invisibly tensored with $\mathbb{Q}$:

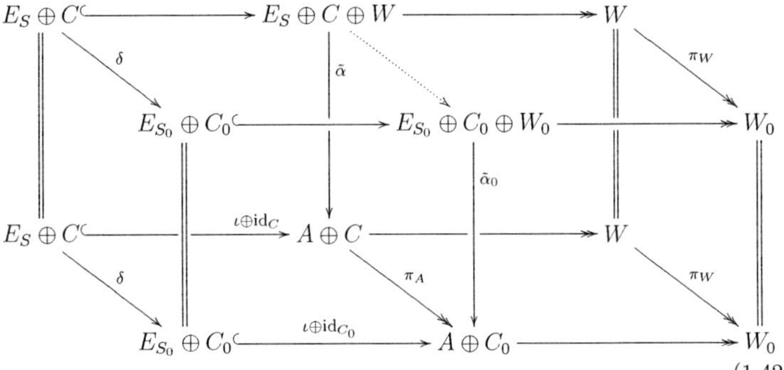

(1.42)

Lemma 1.1.2 again implies that we may choose isomorphisms $\tilde{\alpha}$ and $\tilde{\alpha}_0$ such that the dotted arrow in the diagram above is given by $\delta \oplus \pi_W$. The bottom surface even exists before tensoring with $\mathbb{Q}$. Hence, the Snake Lemma yields an exact sequence

$$\Delta G_{\mathfrak{P}_0} \cdot \mathbb{Z}G \rightarrowtail \mathbb{Z}G \xrightarrow{\pi'} \ker \pi_W \twoheadrightarrow \operatorname{cok} \delta. \tag{1.43}$$

The cokernel $\operatorname{cok} \delta$ is finite, but in general not zero.

Now we can put everything together in the following large commutative diagram which defines an automorphism ψ of $\mathbb{Q}G$.

$$
\begin{array}{ccccc}
\mathbb{Q}G & \hookrightarrow & \mathbb{Q}B & \twoheadrightarrow & \mathbb{Q}B_0 \\
\downarrow{\scriptstyle \tilde{\beta}'} & & \downarrow{\scriptstyle \tilde{\beta}} & & \downarrow{\scriptstyle \tilde{\beta}_0} \\
\operatorname{ind}_{G_{\mathfrak{P}_0}}^{G}\mathbb{Q} \oplus \mathbb{Q}G/N_{G_{\mathfrak{P}_0}} & \hookrightarrow & \mathbb{Q}R \oplus \mathbb{Q}\overline{\nabla}_S & \xrightarrow{\pi_R \oplus \pi_{\overline{\nabla}}} & \mathbb{Q}R_0 \oplus \mathbb{Q}\overline{\nabla}_{S_0} \\
\downarrow{\scriptstyle \mathrm{id}\oplus\phi'} & & \downarrow{\scriptstyle \mathrm{id}\oplus\phi} & & \downarrow{\scriptstyle \mathrm{id}\oplus\phi_0} \\
\operatorname{ind}_{G_{\mathfrak{P}_0}}^{G}\mathbb{Q} \oplus \Delta G_{\mathfrak{P}_0} \cdot \mathbb{Q}G & \hookrightarrow & \mathbb{Q}R \oplus \mathbb{Q}(E_S \oplus C) & \xrightarrow{\pi_R \oplus \delta} & \mathbb{Q}R_0 \oplus \mathbb{Q}(E_{S_0} \oplus C_0) \\
\downarrow & & \downarrow & & \downarrow \\
\mathbb{Q}\ker \pi_W \oplus \Delta G_{\mathfrak{P}_0} \cdot \mathbb{Q}G & \hookrightarrow & \mathbb{Q}W \oplus \mathbb{Q}(E_S \oplus C) & \xrightarrow{\pi_W \oplus \delta} & \mathbb{Q}W_0 \oplus \mathbb{Q}(E_{S_0} \oplus C_0) \\
\downarrow{\scriptstyle \tilde{\alpha}'} & & \downarrow{\scriptstyle \tilde{\alpha}} & & \downarrow{\scriptstyle \tilde{\alpha}_0} \\
\mathbb{Q}G & \hookrightarrow & \mathbb{Q}(A \oplus C) & \twoheadrightarrow & \mathbb{Q}(A \oplus C_0)
\end{array}
$$

(with $\psi : \mathbb{Q}G \to \mathbb{Q}G$ the composite of the left column)

Since the upper and bottom exact sequences already exist at $\mathbb{Z}G$-level, we get

$$
\begin{aligned}
\Omega_{\phi_0} - \Omega_{\phi} &= -(\mathbb{Z}G, \psi, \mathbb{Z}G) \\
&= -\partial[\mathbb{Q}G, \psi].
\end{aligned} \tag{1.44}
$$

To have full knowledge of the automorphism ψ it suffices to compute $\psi(1)$. For this, we have to start with the map $\tilde{\beta}'$ which locally derives from

$$
\begin{array}{ccccc}
\mathbb{Q} & \hookrightarrow & \mathbb{Q}G_{\mathfrak{P}_0} & \twoheadrightarrow & \mathbb{Q}G_{\mathfrak{P}_0}/N_{G_{\mathfrak{P}_0}} \\
\| & & \downarrow{\scriptstyle (\tilde{\beta}')^{\mathrm{loc}}} & & \| \\
\mathbb{Q} & \hookrightarrow & \mathbb{Q} \oplus \mathbb{Q}G_{\mathfrak{P}_0}/N_{G_{\mathfrak{P}_0}} & \twoheadrightarrow & \mathbb{Q}G_{\mathfrak{P}_0}/N_{G_{\mathfrak{P}_0}},
\end{array}
$$

where we again identify $\Delta G^0_{\mathfrak{P}_0}$ with $\mathbb{Z}G_{\mathfrak{P}_0}/N_{G_{\mathfrak{P}_0}}$. By the K_1-Simplification Lemma 1.1.3 we may assume that

$$
(\tilde{\beta}')^{\mathrm{loc}}(1) = (\frac{1}{|G_{\mathfrak{P}_0}|}, 1 \bmod N_{G_{\mathfrak{P}_0}}).
$$

The map ϕ' is already known and we can neglect the inclusion $i : \ker \pi_W \rightarrowtail \operatorname{ind}_{G_{\mathfrak{P}_0}}^{G}\mathbb{Z}$. The map $\tilde{\alpha}'$ derives from the commutative diagram

$$
\begin{array}{ccccc}
\Delta G_{\mathfrak{P}_0} \cdot \mathbb{Q}G & \hookrightarrow & \mathbb{Q}\ker \pi_W \oplus \Delta G_{\mathfrak{P}_0} \cdot \mathbb{Q}G & \twoheadrightarrow & \mathbb{Q}\ker \pi_W \\
\| & & \downarrow{\scriptstyle \tilde{\alpha}'} & & \| \\
\Delta G_{\mathfrak{P}_0} \cdot \mathbb{Q}G & \hookrightarrow & \mathbb{Q}G & \twoheadrightarrow & \mathbb{Q}\ker \pi_W
\end{array}
$$

By Proposition 4.1 of [Gr3] the map $\mathbb{Q}G \to \mathbb{Q}\ker\pi_W$ is multiplication by $-h \cdot \frac{1}{|G_{\mathfrak{P}_0}|} N_{G_{\mathfrak{P}_0}}$. To motivate this a little, note that we surely have to multiply by the idempotent $\varepsilon_0 := \frac{1}{|G_{\mathfrak{P}_0}|} N_{G_{\mathfrak{P}_0}}$ since $\ker\pi_W \subset \mathrm{ind}\,_{G_{\mathfrak{P}_0}}^G \mathbb{Z}$. Moreover, h annihilates $\mathrm{cok}\,\delta$ (cf. sequence (1.43)).

Again by K_1-Simplification (Lemma 1.1.3) we may assume that

$$\tilde{\alpha}'(x \otimes 1, y) = y - h^{-1}\varepsilon_0 \cdot x,$$

where $x \otimes 1 \in \ker\pi_W \subset \mathrm{ind}\,_{G_{\mathfrak{P}_0}}^G \mathbb{Z} = \mathbb{Z}G \otimes_{\mathbb{Z}G_{\mathfrak{P}_0}} \mathbb{Z}$ and $y \in \Delta G_{\mathfrak{P}_0} \cdot \mathbb{Z}G$.

We compute

$$\begin{aligned}\psi(1) &= \tilde{\alpha}'(i^{-1} \oplus \mathrm{id})(\mathrm{id} \oplus \phi')\tilde{\beta}'(1) \\ &= \tilde{\alpha}'(i^{-1} \oplus \mathrm{id})(\mathrm{id} \oplus \phi')(\tfrac{1}{|G_{\mathfrak{P}_0}|} \otimes 1, 1 \bmod N_{G_{\mathfrak{P}_0}}) \\ &= \tilde{\alpha}'(i^{-1} \oplus \mathrm{id})(\tfrac{1}{|G_{\mathfrak{P}_0}|} \otimes 1, 1 - \varepsilon_0) \\ &= 1 - \varepsilon_0 - \tfrac{1}{h|G_{\mathfrak{P}_0}|}\varepsilon_0.\end{aligned}$$

Therefore, we get

$$\Omega_{\phi_0} - \Omega_\phi = \partial[\mathrm{ind}\,_{G_{\mathfrak{P}_0}}^G \mathbb{Q}, -h|G_{\mathfrak{P}_0}|]$$

by (1.44). This proves Theorem 1.4.3. □

To complete this paragraph, we have to discuss how Ω_ϕ varies if S is enlarged by the orbit of a non-ramified prime $\mathfrak{P}_0$. As before let $S_0 := S \cup G\mathfrak{P}_0$. The exact sequence (1.13) for the sets S and S_0 together with the natural exact sequence $\mathbb{Z}S \rightarrowtail \mathbb{Z}S_0 \twoheadrightarrow \mathrm{ind}\,_{G_{\mathfrak{P}_0}}^G \mathbb{Z}$ yield an exact sequence

$$\overline{\nabla}_S \rightarrowtail \overline{\nabla}_{S_0} \twoheadrightarrow \mathrm{ind}\,_{G_{\mathfrak{P}_0}}^G \mathbb{Z}.$$

On the other hand, the map

$$E_{S_0} \to \mathbb{Z}[G/G_{\mathfrak{P}_0}] = \mathrm{ind}\,_{G_{\mathfrak{P}_0}}^G \mathbb{Z},\ u \mapsto \sum_{g \in G/G_{\mathfrak{P}_0}} v_{\mathfrak{P}_0}(u^g)g$$

has kernel E_S and finite cokernel. Thus, for each isomorphism $\phi : \mathbb{Q}\overline{\nabla}_S \to \mathbb{Q}(E_S \oplus C)$, where C is $\mathbb{Z}G$-free of rank $|S_{\mathrm{ram}}^* \setminus (S \cap S_{\mathrm{ram}})^*|$, there is an isomorphism ϕ_0 fitting in a commutative diagram

$$\begin{array}{ccc} \mathbb{Q}\overline{\nabla}_S & \xrightarrow{\phi} & \mathbb{Q}(E_S \oplus C) \\ \downarrow & & \downarrow \\ \mathbb{Q}\overline{\nabla}_{S_0} & \xrightarrow{\phi_0} & \mathbb{Q}(E_{S_0} \oplus C) \\ \downarrow & & \downarrow \\ \operatorname{ind}_{G_{\mathfrak{P}_0}}^{G} \mathbb{Q} & = & \operatorname{ind}_{G_{\mathfrak{P}_0}}^{G} \mathbb{Q} \end{array} \tag{1.45}$$

The result corresponding to Theorem 1.4.3 is exactly the same as for large sets S (cf. [GRW], p. 60):

Theorem 1.4.4 *Fix a set of data (D) and let $\mathfrak{P}_0$ be a prime not in S which does not ramify in L/K. Given a $\mathbb{Q}G$-isomorphism ϕ_0 that fits in diagram (1.45) we have an equality*

$$\Omega_{\phi_0} - \Omega_\phi = \partial[\mathbb{Q}G, \eta].$$

Here, η ist the $\mathbb{Q}G$-automorphism given by

$$\eta(1) = |G_{\mathfrak{P}_0}|\varepsilon_0 + \frac{1}{|G_{\mathfrak{P}_0}|} \sum_{i=0}^{|G_{\mathfrak{P}_0}|-1} i\phi_{\mathfrak{P}_0}^i (1 - \varepsilon_0),$$

where $\varepsilon_0 = \frac{1}{|G_{\mathfrak{P}_0}|} N_{G_{\mathfrak{P}_0}}$ and $\phi_{\mathfrak{P}_0}$ is the Frobenius automorphism at $\mathfrak{P}_0$. In particular, $\Omega_{\phi_0} - \Omega_\phi$ has representing homomorphism

$$\chi \mapsto (|G_{\mathfrak{P}_0}|)^{\dim V_\chi^{G_{\mathfrak{P}_0}}} \cdot \det(\phi_{\mathfrak{P}_0} - 1 | V_\chi / V_\chi^{G_{\mathfrak{P}_0}})^{-1},$$

where V_χ is a $\mathbb{C}G$-module with character χ.

Proof. Due to Theorem 1.4.3 and Proposition 1.4.2 we may assume that S (and thus also S_0) contains all the ramified primes. Hence, $\overline{\nabla}_S = \Delta S$ and $\overline{\nabla}_{S_0} = \Delta S_0$.
As before let S' be a finite set of places of L which contains $S = S \cup S_{\mathrm{ram}}$ and is large enough to generate the ideal class group of L, and such that $\bigcup_{\mathfrak{P} \in S'} G_{\mathfrak{P}} = G$. But this time we insist in the additional property that $\mathfrak{P}_0 \notin S'$ and set $S_0' := S' \cup G\mathfrak{P}_0$. The first step in the construction of Tate sequences then gives rise to the commutative diagram

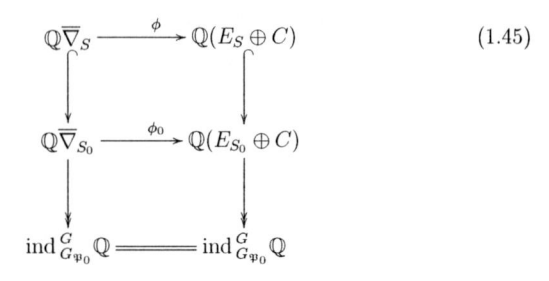

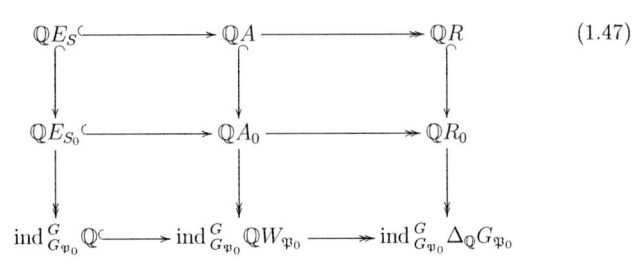

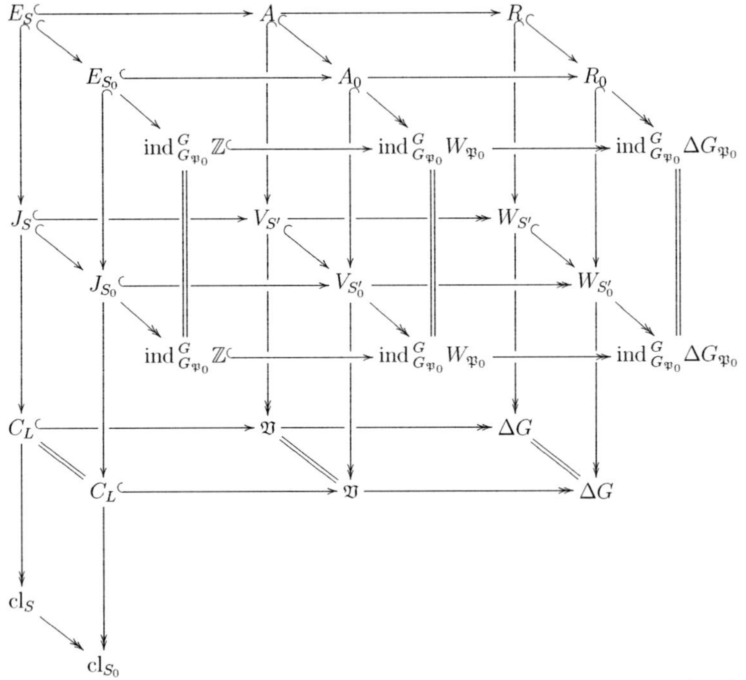

(1.46)

Recall that $W_{\mathfrak{P}_0} \subset \Delta G_{\mathfrak{P}_0} \times \mathbb{Z}\overline{G_{\mathfrak{P}_0}} = \Delta G_{\mathfrak{P}_0} \times \mathbb{Z}G_{\mathfrak{P}_0}$ since $\mathfrak{P}_0$ is unramified in L/K. The projection to the second component induces an isomorphism $\mathrm{pr}_y : W_{\mathfrak{P}_0} \simeq \mathbb{Z}G_{\mathfrak{P}_0}$. Hence, the sequence

$$A \rightarrowtail A_0 \twoheadrightarrow \operatorname{ind}_{G_{\mathfrak{P}_0}}^G W_{\mathfrak{P}_0}$$

is an exact sequence of c.t. $\mathbb{Z}G$-modules. Furthermore, the roof of the above diagram consists of exact rows and columns after tensoring with $\mathbb{Q}$:

$$\begin{array}{ccccc}
\mathbb{Q}E_S & \hookrightarrow & \mathbb{Q}A & \twoheadrightarrow & \mathbb{Q}R \\
\downarrow & & \downarrow & & \downarrow \\
\mathbb{Q}E_{S_0} & \hookrightarrow & \mathbb{Q}A_0 & \twoheadrightarrow & \mathbb{Q}R_0 \\
\downarrow & & \downarrow & & \downarrow \\
\operatorname{ind}_{G_{\mathfrak{P}_0}}^G \mathbb{Q} & \hookrightarrow & \operatorname{ind}_{G_{\mathfrak{P}_0}}^G \mathbb{Q}W_{\mathfrak{P}_0} & \twoheadrightarrow & \operatorname{ind}_{G_{\mathfrak{P}_0}}^G \Delta_{\mathbb{Q}}G_{\mathfrak{P}_0}
\end{array} \qquad (1.47)$$

If we identify $\operatorname{ind}_{G_{\mathfrak{P}_0}}^G \mathbb{Q}W_{\mathfrak{P}_0}$ with $\operatorname{ind}_{G_{\mathfrak{P}_0}}^G \mathbb{Q}G_{\mathfrak{P}_0}$, the injection of the bottom sequence in (1.47) is induced by $1 \mapsto N_{G_{\mathfrak{P}_0}}$.

The second step in the construction of Tate sequences yields a commutative diagram

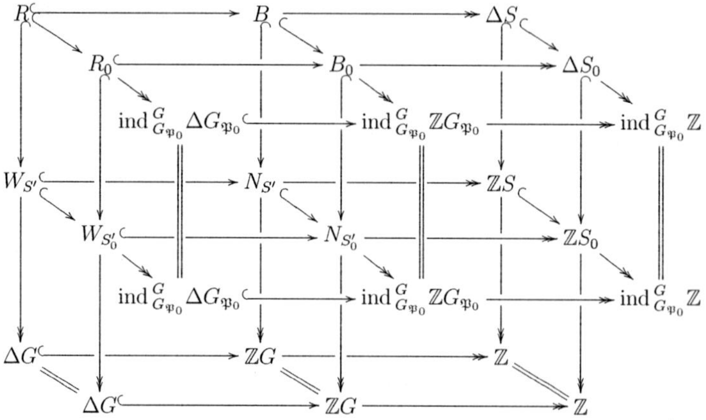

(1.48)

As before, we choose the automorphisms β of $\mathbb{Q}R$ and β_0 of $\mathbb{Q}R_0$ to be the identity. We get a diagram whose top is that of diagram (1.48) tensored with $\mathbb{Q}$:

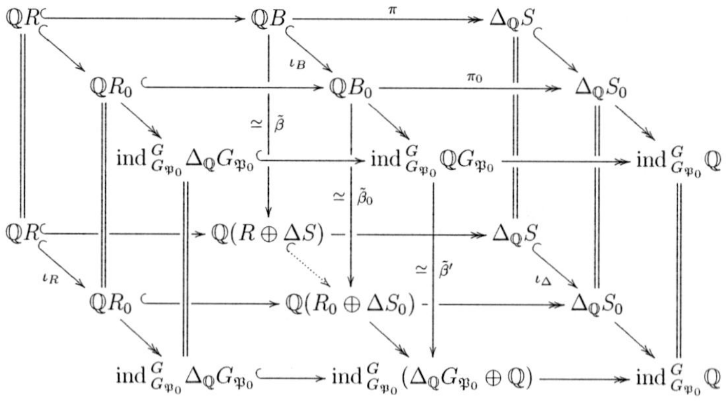

As on earlier occasions, Lemma 1.1.2 implies that we can choose $\tilde{\beta}$ and $\tilde{\beta}_0$ such that the dotted injection $\mathbb{Q}(R \oplus \Delta S) \rightarrowtail \mathbb{Q}(R_0 \oplus \Delta S_0)$ in the above diagram is $\iota_R \oplus \iota_\Delta$.

We also choose the automorphisms α and α_0 to be the identity and get a diagram

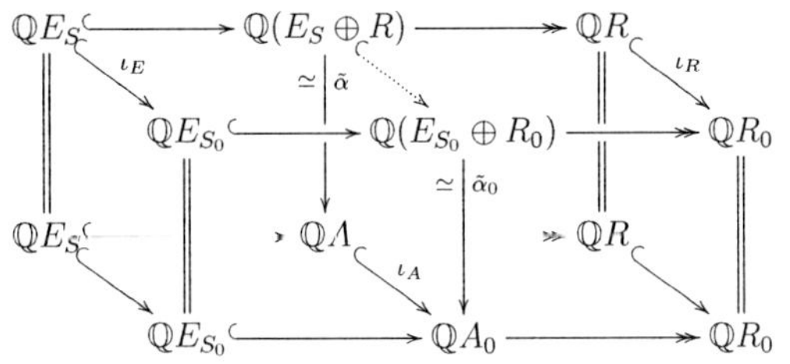

in which the maps $\tilde\alpha$ and $\tilde\alpha_0$ are taken via Lemma 1.1.2 such that the dotted arrow is just $\iota_E \oplus \iota_R$.

Putting things together, we get the following commutative diagram which defines an isomorphism η:

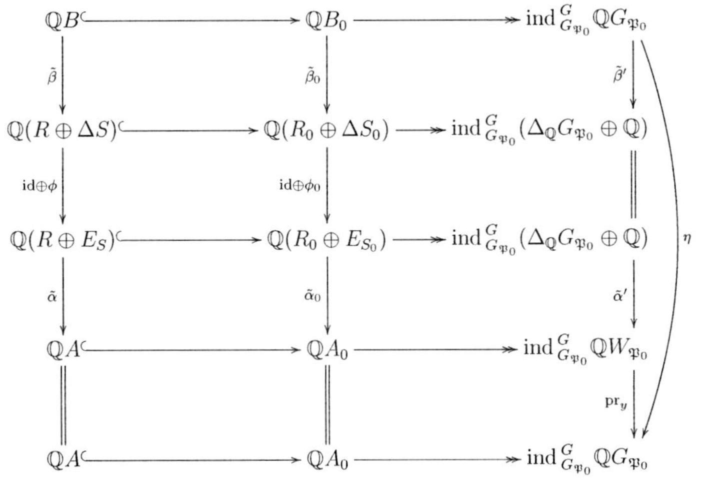

Note that the upper and bottom sequence already exist at $\mathbb{Z}G$-level, and so does the isomorphism pr_y. Hence, by Lemma 1.1.6

$$\Omega_{\phi_0} - \Omega_\phi = (\mathbb{Z}G, \eta, \mathbb{Z}G) = \partial[\mathbb{Q}G, \eta].$$

We are left with the computation of $\eta(1)$. By the K_1-simplification Lemma 1.1.3 and the definition of $\tilde\beta'$ we may assume that

$$\tilde\beta'(1) = (1 - \varepsilon_0, 1).$$

The isomorphism $\mathrm{pr}_y \circ \tilde{\alpha}'$ fits into a diagram

$$\begin{array}{ccccc}
\mathrm{ind}_{G_{\mathfrak{P}_0}}^{G} \mathbb{Q} & \hookrightarrow & \mathrm{ind}_{G_{\mathfrak{P}_0}}^{G} (\Delta_{\mathbb{Q}} G_{\mathfrak{P}_0} \oplus \mathbb{Q}) & \twoheadrightarrow & \mathrm{ind}_{G_{\mathfrak{P}_0}}^{G} \Delta_{\mathbb{Q}} G_{\mathfrak{P}_0} \\
\| & & \downarrow{\scriptstyle \mathrm{pr}_y \tilde{\alpha}'} & & \| \\
\mathrm{ind}_{G_{\mathfrak{P}_0}}^{G} \mathbb{Q} & \xrightarrow{1 \mapsto N_{G_{\mathfrak{P}_0}}} & \mathrm{ind}_{G_{\mathfrak{P}_0}}^{G} \mathbb{Q} G_{\mathfrak{P}_0} & \xrightarrow{1 \mapsto \phi_{\mathfrak{P}_0} - 1} & \mathrm{ind}_{G_{\mathfrak{P}_0}}^{G} \Delta_{\mathbb{Q}} G_{\mathfrak{P}_0}
\end{array}$$

and again by K_1-simplification we may assume that

$$\mathrm{pr}_y \tilde{\alpha}'(x,q) = N_{G_{\mathfrak{P}_0}} q + (1-\varepsilon_0) \left(\frac{1}{|G_{\mathfrak{P}_0}|} \sum_{i=0}^{|G_{\mathfrak{P}_0}|-1} i \phi_{\mathfrak{P}_0}^i \right) x.$$

Hence, we finally get

$$\begin{aligned}
\eta(1) &= \mathrm{pr}_y \tilde{\alpha}' \tilde{\beta}'(1) \\
&= \mathrm{pr}_y \tilde{\alpha}'(1-\varepsilon_0, 1) \\
&= N_{G_{\mathfrak{P}_0}} + (1-\varepsilon_0) \tfrac{1}{|G_{\mathfrak{P}_0}|} \textstyle\sum_{i=0}^{|G_{\mathfrak{P}_0}|-1} i \phi_{\mathfrak{P}_0}^i
\end{aligned}$$

as desired. □

1.5 The conjecture

Thanks to the results of the last paragraph we are now able to state the LRNC for small sets of places. But before doing so we recall the basic ingredients of this conjecture apart from the element Ω_ϕ.

So let us fix a finite Galois extension L/K of number fields with Galois group G and a finite G-invariant set S of places of L, which contains all the archimedean primes. Then there are $\mathbb{Q}G$-isomorphisms

$$\phi : \Delta_{\mathbb{Q}} S \xrightarrow{\simeq} \mathbb{Q} E_S,$$

and the Stark-Tate regulator is defined to be

$$\begin{aligned}
R_\phi : R(G) &\rightarrow \mathbb{C}^\times \\
\chi &\mapsto \det(\lambda_S \phi | \mathrm{Hom}_G(V_{\check{\chi}}, \Delta_{\mathbb{C}} S)),
\end{aligned}$$

where λ_S is the Dirichlet map (1.14) and $V_{\check{\chi}}$ is a $\mathbb{C}G$-module whose character is contragredient to χ. Furthermore, let $S(K) := \{\mathfrak{P} \cap K | \mathfrak{P} \in S\}$ and $c_S(\chi)$ be the leading coefficient of the Taylor expansion of the S-truncated L-function $L_S(L/K,\chi,s)$ at $s=0$. For $\Re(s) > 1$ this is the function

$$L_S(L/K,\chi,s) = \prod_{\mathfrak{p} \notin S(K)} \det(1 - \phi_{\mathfrak{P}} N(\mathfrak{p})^{-s} | V_\chi^{I_{\mathfrak{P}}}).$$

One defines

$$\begin{array}{rcl} A_\phi : R(G) & \rightarrow & \mathbb{C}^\times \\ \chi & \mapsto & R_\phi(\chi)/c_S(\chi). \end{array}$$

If we fix an algebraic closure $\mathbb{Q}^c$ of $\mathbb{Q}$, there is the following conjecture of Stark:

Conjecture 1.5.1 (Stark) *$A_\phi(\chi^\sigma) = A_\phi(\chi)^\sigma$ for all $\sigma \in \mathrm{Gal}(\mathbb{Q}^c/\mathbb{Q})$.*

Alternatively, one can choose a number field F, Galois over $\mathbb{Q}$ with Galois group Γ, which is large enough such that all representations of G can be realized over F. Then conjecture 1.5.1 is equivalent to $A_\phi(\chi^\sigma) = A_\phi(\chi)^\sigma$ for all $\sigma \in \Gamma$, i.e. $A_\phi \in \mathrm{Hom}_\Gamma(R(G), F^\times)$.

Let us denote by $W(\chi)$ the Artin root number of the character χ. Then it holds (cf. [We], Prop. 7(b), p.57):

Proposition 1.5.2 *If χ is an irreducible symplectic character of G, then $A_\phi(\chi)W(\chi) \in \mathbb{R}^+$.*

Now we fix an embedding $F \rightarrowtail \mathbb{C}$ and denote the corresponding infinite prime by $\wp_\infty$. Define $W(L/K, \cdot) \in \mathrm{Hom}_\Gamma(R(G), J_F)$ by

$$W(L/K,\chi)_\wp = \begin{cases} W(\chi^{\gamma^{-1}}) & \text{if } \chi \text{ is symplectic and } \wp = \wp_\infty^\gamma \\ 1 & \text{otherwise} \end{cases}$$

such that the homomorphism $\chi \mapsto A_\phi(\chi)W(L/K,\chi)$ is in $\mathrm{Hom}_\Gamma^+(R(G), J_F)$ if Stark's conjecture holds.

For large S the LRNC now states

Conjecture 1.5.3 (LRNC for large S) *The element $\Omega_\phi \in K_0T(\mathbb{Z}G)$ has representing homomorphism $\chi \mapsto A_\phi(\check{\chi})W(L/K,\check{\chi})$.*

In the construction of Ω_ϕ for small sets S, the module ΔS has been replaced by $\overline{\nabla}_S$. We aim to define a modified Dirichlet map

$$\lambda_S^{\mathrm{mod}} : E_S \oplus C \longrightarrow \mathbb{R} \otimes \overline{\nabla}_S,$$

where C is a free $\mathbb{Z}G$-module of rank $|S^*_{\mathrm{ram}} \setminus (S \cap S_{\mathrm{ram}})^*|$. For this, we have to take a closer look at the modules $W^0_{\mathfrak{P}}$, especially for ramified primes $\mathfrak{P}$.

Let us write $\phi_{\mathfrak{P}}$ for the Frobenius automorphism at $\mathfrak{P}$ as well as for a fixed lift of it. Recall the definition of the inertial lattice (cf. (1.11))

$$W_{\mathfrak{P}} = \{(x,y) \in \Delta G_{\mathfrak{P}} \oplus \mathbb{Z}\overline{G_{\mathfrak{P}}} : \overline{x} = (\phi_{\mathfrak{P}} - 1)y\}.$$

Obviously, $W_{\mathfrak{P}}$ is the kernel of the map

$$\begin{array}{rcl} \Delta G_{\mathfrak{P}} \times \mathbb{Z}\overline{G_{\mathfrak{P}}} & \longrightarrow & \mathbb{Z}\overline{G_{\mathfrak{P}}} \\ (g-1, \overline{h}) & \mapsto & \overline{g} - 1 + (1-\phi_{\mathfrak{P}})\overline{h}. \end{array}$$

Hence, using the identifications concerning $\mathbb{Z}$-duals explained in the preliminaries, we achieve a description of $W^0_{\mathfrak{P}}$ as the cokernel of the map (cf. [Gr3], p.20)

$$\begin{array}{rcl} \mathbb{Z}\overline{G_{\mathfrak{P}}} & \longrightarrow & \mathbb{Z}G_{\mathfrak{P}}/N_{G_{\mathfrak{P}}} \times \mathbb{Z}\overline{G_{\mathfrak{P}}} \\ 1 & \mapsto & (N_{I_{\mathfrak{P}}}, 1-\phi_{\mathfrak{P}}^{-1}). \end{array}$$

Proposition 1.5.4 *Let κ denote the canonical epimorphism from $\mathbb{Z}G^2_{\mathfrak{P}}$ onto $W^0_{\mathfrak{P}}$ and define*

$$\begin{array}{rcl} q : W_{\mathfrak{P}} & \longrightarrow & \mathbb{Z}G^2_{\mathfrak{P}} \\ (x,y) & \mapsto & (N_{I_{\mathfrak{P}}}y, \phi_{\mathfrak{P}}^{-1}x). \end{array}$$

Then it holds:

(1) The kernel of κ is generated by $z = (N_{I_{\mathfrak{P}}}, 1-\phi_{\mathfrak{P}}^{-1})$ and $0 \times \Delta(G_{\mathfrak{P}}, \overline{G_{\mathfrak{P}}})$, where $\Delta(G_{\mathfrak{P}}, \overline{G_{\mathfrak{P}}})$ is the kernel of the canonical projection $\mathbb{Z}G_{\mathfrak{P}} \twoheadrightarrow \mathbb{Z}\overline{G_{\mathfrak{P}}}$.

(2) The diagram

$$\begin{array}{ccccc} \mathbb{Z} & \xrightarrow{1\mapsto N_{G_{\mathfrak{P}}}} & \mathbb{Z}G_{\mathfrak{P}} & \twoheadrightarrow & \mathbb{Z}G_{\mathfrak{P}}/N_{G_{\mathfrak{P}}} \\ \downarrow & & \downarrow{\scriptstyle \iota_1} & & \downarrow{\scriptstyle \iota_1} \\ W_{\mathfrak{P}} & \xrightarrow{q} & \mathbb{Z}G^2_{\mathfrak{P}} & \xrightarrow{\kappa} & W^0_{\mathfrak{P}} \\ \downarrow{\scriptstyle \mathrm{pr}_x} & & \downarrow{\scriptstyle \phi_{\mathfrak{P}}\cdot \mathrm{pr}_2} & & \downarrow{\scriptstyle (0,\mathrm{aug}_{\overline{G_{\mathfrak{P}}}})} \\ \Delta G_{\mathfrak{P}} & \hookrightarrow & \mathbb{Z}G_{\mathfrak{P}} & \twoheadrightarrow & \mathbb{Z} \end{array}$$

commutes and has exact rows and columns.

PROOF. The diagram is taken from [GW], Lemma 4.1, but see [Gr3], p.20 et seq., where the full proposition is proved and wherefrom we have adopted most of the notation. □

We now set

$$d_{\mathfrak{P}} := \frac{1}{|G_{\mathfrak{P}}|}\kappa(|G_{\mathfrak{P}}|, N_{G_{\mathfrak{P}}}) \in \mathbb{Q}W^0_{\mathfrak{P}}.$$

Observe that this definition differs from the corresponding element $d_{\mathfrak{p}}$ in [Gr3].

Lemma 1.5.5 *$d_{\mathfrak{P}}$ is a $\mathbb{Q}G_{\mathfrak{P}}$-generator of $\mathbb{Q}W^0_{\mathfrak{P}}$.*

PROOF. It suffices to show that $\kappa(0,1) \in \mathbb{Q}G_{\mathfrak{P}} \cdot d_{\mathfrak{P}}$, since in this case also $\kappa(1,0) \in \mathbb{Q}G_{\mathfrak{P}} \cdot d_{\mathfrak{P}}$ and these two generate $W^0_{\mathfrak{P}}$. Let us set $e_{\mathfrak{P}} = |I_{\mathfrak{P}}|$ and $f_{\mathfrak{P}} = |\overline{G_{\mathfrak{P}}}|$. By means of Proposition 1.5.4 we may compute

$$\begin{array}{rcl} N_{I_{\mathfrak{P}}} d_{\mathfrak{P}} & = & \kappa(N_{I_{\mathfrak{P}}}, f_{\mathfrak{P}}^{-1}N_{G_{\mathfrak{P}}}) \\ & = & \kappa(0, f_{\mathfrak{P}}^{-1}N_{G_{\mathfrak{P}}} + \phi_{\mathfrak{P}}^{-1} - 1) \\ & = & \kappa(0, f_{\mathfrak{P}}^{-1}N_{G_{\mathfrak{P}}} + (\phi_{\mathfrak{P}}^{-1} - 1)e_{\mathfrak{P}}^{-1}N_{I_{\mathfrak{P}}} + 1 - e_{\mathfrak{P}}^{-1}N_{I_{\mathfrak{P}}}) \\ & = & h_{\mathfrak{P}}\kappa(0,1), \end{array}$$

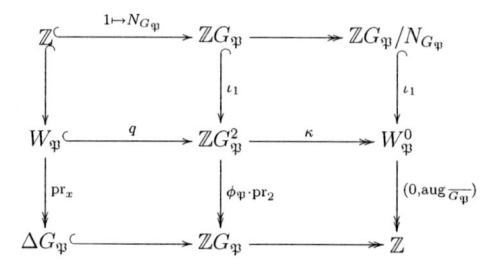

where $h_{\mathfrak{P}} = f_{\mathfrak{P}}^{-1} N_{G_{\mathfrak{P}}} + (\phi_{\mathfrak{P}}^{-1} - 1) e_{\mathfrak{P}}^{-1} N_{I_{\mathfrak{P}}} + 1 - e_{\mathfrak{P}}^{-1} N_{I_{\mathfrak{P}}} \in \mathbb{Q}G_{\mathfrak{P}}^{\times}$. Indeed, if we decompose 1 into central idempotents, namely

$$1 = |G_{\mathfrak{P}}|^{-1} N_{G_{\mathfrak{P}}} + e_{\mathfrak{P}}^{-1} N_{I_{\mathfrak{P}}} (1 - |G_{\mathfrak{P}}|^{-1} N_{G_{\mathfrak{P}}}) + 1 - e_{\mathfrak{P}}^{-1} N_{I_{\mathfrak{P}}},$$

we find out that

$$h_{\mathfrak{P}}^{-1} = e_{\mathfrak{P}}^{-1} |G_{\mathfrak{P}}|^{-1} N_{G_{\mathfrak{P}}} + f_{\mathfrak{P}}^{-1} \sum_{i=0}^{f_{\mathfrak{P}}-1} i \phi_{\mathfrak{P}}^{-i} e_{\mathfrak{P}}^{-1} N_{I_{\mathfrak{P}}} (1 - |G_{\mathfrak{P}}|^{-1} N_{G_{\mathfrak{P}}}) + 1 - e_{\mathfrak{P}}^{-1} N_{I_{\mathfrak{P}}}.$$

Hence, $\kappa(0,1) = h_{\mathfrak{P}}^{-1} N_{I_{\mathfrak{P}}} d_{\mathfrak{P}} \in \mathbb{Q}G_{\mathfrak{P}} \cdot d_{\mathfrak{P}}$. □

Let $1_{\mathfrak{P}}$, $\mathfrak{P} \in S_{\mathrm{ram}}^* \setminus (S \cap S_{\mathrm{ram}})^*$ be a $\mathbb{Z}G$-basis of the free $\mathbb{Z}G$-module C. We choose a positive multiple h of h_L, the class number of L, and $u_{\mathfrak{P}} \in L$ such that $v_{\mathfrak{P}}(u_{\mathfrak{P}}) = h$ and $v_{\mathfrak{Q}}(u_{\mathfrak{P}}) = 0$ for all finite primes $\mathfrak{Q} \neq \mathfrak{P}$. We define

$$\begin{aligned} \lambda_C : C &\longrightarrow \mathbb{R} \otimes \bigoplus_{\mathfrak{P} \in S_{\mathrm{ram}}^* \setminus (S \cap S_{\mathrm{ram}})^*} \mathrm{ind}_{G_{\mathfrak{P}}}^G W_{\mathfrak{P}}^0 \oplus \mathbb{R} S_\infty \\ 1_{\mathfrak{P}} &\mapsto \left(h \log N(\mathfrak{P}) \frac{1}{|G_{\mathfrak{P}}|} N_{G_{\mathfrak{P}}} + 1 - \frac{1}{|G_{\mathfrak{P}}|} N_{G_{\mathfrak{P}}} \right) d_{\mathfrak{P}} - \sum_{\mathfrak{Q} | \infty} \log |u_{\mathfrak{P}}|_{\mathfrak{Q}} \mathfrak{Q}. \end{aligned}$$

By the second part of Proposition 1.5.4 we have

$$(0, \mathrm{aug}_{\overline{G_{\mathfrak{P}}}})(d_{\mathfrak{P}}) = \mathrm{aug}\,(\phi_{\mathfrak{P}} \mathrm{pr}_2(1, \frac{1}{|G_{\mathfrak{P}}|} N_{G_{\mathfrak{P}}})) = 1.$$

Hence, the projection in sequence (1.13) sends $\lambda_C(1_{\mathfrak{P}})$ to

$$h \log N(\mathfrak{P}) - \sum_{\mathfrak{Q}|\infty} \log |u_{\mathfrak{P}}|_{\mathfrak{Q}} = - \sum_{\text{all } \mathfrak{Q}} \log |u_{\mathfrak{P}}|_{\mathfrak{Q}} = 0.$$

Thus, the image of λ_C lies in $\mathbb{R}\overline{\nabla}$, and we may define a modified Dirichlet map by

$$\begin{aligned} \lambda_S^{\mathrm{mod}} : E_S \oplus C &\longrightarrow \mathbb{R}\overline{\nabla} \\ (e, c) &\mapsto \lambda_S(e) + \lambda_C(c), \end{aligned} \tag{1.49}$$

where λ_S is the usual Dirichlet map (1.14). Note that λ_S^{mod} depends on the choices of h and the elements $u_{\mathfrak{P}}$.

Definition 1.5.6 *We call the map*

$$\begin{aligned} R_\phi^{\mathrm{mod}} : R(G) &\longrightarrow \mathbb{C}^\times \\ \chi &\mapsto \frac{\det(\lambda_S^{\mathrm{mod}} \phi | \mathrm{Hom}_G(V_{\check{\chi}}, \mathbb{C}\overline{\nabla}_S))}{\prod_{\mathfrak{P} \in S_{\mathrm{ram}}^* \setminus (S \cap S_{\mathrm{ram}})^*} (-h|G_{\mathfrak{P}}|)^{\dim V_{\check{\chi}}^{G_{\mathfrak{P}}}}} \end{aligned}$$

the ***modified Stark-Tate regulator*** *and set*

$$\begin{aligned} A_\phi^{\mathrm{mod}} : R(G) &\longrightarrow \mathbb{C}^\times \\ \chi &\mapsto \frac{R_\phi^{\mathrm{mod}}(\chi)}{c_{S \cup S_{\mathrm{ram}}}(\chi)}. \end{aligned}$$

REMARK. If the set S already contains all the ramified primes, we obviously have $R_\phi^{\mathrm{mod}} = R_\phi$ and $A_\phi^{\mathrm{mod}} = A_\phi$.

Unfortunately, the above definition is not independent of the choices of h and the $u_{\mathfrak{P}}$. Nevertheless, we have the following

Proposition 1.5.7 *The maps* R_ϕ^{mod}, $A_\phi^{\mathrm{mod}} \in \mathrm{Hom}(R(G), \mathbb{C}^\times)$ *are well defined modulo* $\mathrm{Det}\,(U(\mathbb{Z}G))$.

PROOF. Let $\mathfrak{P} \in S_{\mathrm{ram}}^* \setminus (S \cap S_{\mathrm{ram}})^*$ and assume that we have defined $\tilde{\lambda}_S^{\mathrm{mod}}$ by another choice $\tilde{u}_{\mathfrak{P}} \in L^\times$ such that $v_{\mathfrak{P}}(\tilde{u}_{\mathfrak{P}}) = h$ and $v_{\mathfrak{Q}}(\tilde{u}_{\mathfrak{P}}) = 0$ for all finite primes $\mathfrak{Q} \neq \mathfrak{P}$. Then there is a unit $e_{\mathfrak{P}} \in \mathfrak{o}_L^\times$ with the property that $e_{\mathfrak{P}}\tilde{u}_{\mathfrak{P}} = u_{\mathfrak{P}}$. We have a commutative square

$$\begin{array}{ccc} E_S \oplus C & \xrightarrow{\lambda_S^{\mathrm{mod}}} & \mathbb{C}\overline{\nabla}_S \\ \downarrow \psi & & \| \\ E_S \oplus C & \xrightarrow{\tilde{\lambda}_S^{\mathrm{mod}}} & \mathbb{C}\overline{\nabla}_S \end{array}$$

where ψ is the $\mathbb{Z}G$-automorphism which is the identity on E_S and maps $1_{\mathfrak{P}}$ to $(e_{\mathfrak{P}}, 1_{\mathfrak{P}}) \in E_S \oplus C$. Indeed,

$$\begin{aligned} \tilde{\lambda}_S^{\mathrm{mod}}(e_{\mathfrak{P}}, 1_{\mathfrak{P}}) &= -\sum_{\mathrm{all}\ \mathfrak{Q}} \log|e_{\mathfrak{P}}|_{\mathfrak{Q}}\mathfrak{Q} + d'_{\mathfrak{P}} - \sum_{\mathfrak{Q}|\infty} \log|\tilde{u}_{\mathfrak{P}}|_{\mathfrak{Q}}\mathfrak{Q} \\ &= d'_{\mathfrak{P}} - \sum_{\mathfrak{Q}|\infty} \log|e_{\mathfrak{P}}\tilde{u}_{\mathfrak{P}}|_{\mathfrak{Q}}\mathfrak{Q} \\ &= d'_{\mathfrak{P}} - \sum_{\mathfrak{Q}|\infty} \log|u_{\mathfrak{P}}|_{\mathfrak{Q}}\mathfrak{Q} \\ &= \lambda_S^{\mathrm{mod}}(1, 1_{\mathfrak{P}}), \end{aligned}$$

where $d'_{\mathfrak{P}} = \left(h \log N(\mathfrak{P}) \frac{1}{|G_{\mathfrak{P}}|} N_{G_{\mathfrak{P}}} + 1 - \frac{1}{|G_{\mathfrak{P}}|} N_{G_{\mathfrak{P}}}\right) d_{\mathfrak{P}}$. Thus

$$\begin{aligned} \frac{\det(\lambda_S^{\mathrm{mod}}\phi|\mathrm{Hom}_G(V_{\check{\chi}}, \mathbb{C}\overline{\nabla}_S))}{\det(\tilde{\lambda}_S^{\mathrm{mod}}\phi|\mathrm{Hom}_G(V_{\check{\chi}}, \mathbb{C}\overline{\nabla}_S))} &= \det(\lambda_S^{\mathrm{mod}}(\tilde{\lambda}_S^{\mathrm{mod}})^{-1}|\mathrm{Hom}_G(V_{\check{\chi}}, \mathbb{C}\overline{\nabla}_S)) \\ &= \det(\psi|\mathrm{Hom}_G(V_{\check{\chi}}, \mathbb{C}(E_S \oplus C))), \end{aligned}$$

and the map $\chi \mapsto \det(\psi|\mathrm{Hom}_G(V_\chi, \mathbb{C}(E_S \oplus C)))$ is the representing homomorphism of $\partial[\mathbb{Q}(E_S \oplus C), \mathbb{Q} \otimes \psi]$ and lies in $\mathrm{Det}\,(U(\mathbb{Z}G))$, since ψ already exists at $\mathbb{Z}G$-level (cf. (1.9)).

For the dependance on the integer h, suppose that we have made another choice $\tilde{h}$ to define $\tilde{\lambda}_S^{\mathrm{mod}}$. We may assume that $h \mid \tilde{h}$ and even that $|G_{\mathfrak{P}}|$ divides

$m = \tilde{h}/h$. Write $\varepsilon_{\mathfrak{P}}$ for the idempotent $\frac{1}{|G_{\mathfrak{P}}|}N_{G_{\mathfrak{P}}}$ and choose the $\tilde{u}_{\mathfrak{P}}$ to be $u_{\mathfrak{P}}^{m\varepsilon_{\mathfrak{P}}}$. As verified below, we have a commutative square

$$\begin{array}{ccc} \mathbb{Q}(E_S \oplus C) & \xrightarrow{\tilde{\lambda}_S^{\mathrm{mod}}} & \mathbb{C}\overline{\nabla}_S \\ \downarrow \psi & & \| \\ \mathbb{Q}(E_S \oplus C) & \xrightarrow{\lambda_S^{\mathrm{mod}}} & \mathbb{C}\overline{\nabla}_S \end{array}$$

where the $\mathbb{Q}G$-automorphism ψ is the identity on $\mathbb{Q}E_S$ and is given on $\mathbb{Q}C$ by

$$1_{\mathfrak{P}} \mapsto (u_{\mathfrak{P}}^{\varepsilon_{\mathfrak{P}}-1}, (m\varepsilon_{\mathfrak{P}} + 1 - \varepsilon_{\mathfrak{P}})1_{\mathfrak{P}}).$$

For the commutativity we compute

$$\begin{aligned} \lambda_S^{\mathrm{mod}}(\psi(1_{\mathfrak{P}})) &= \lambda_S^{\mathrm{mod}}(u_{\mathfrak{P}}^{\varepsilon_{\mathfrak{P}}-1}, (m\varepsilon_{\mathfrak{P}} + 1 - \varepsilon_{\mathfrak{P}})1_{\mathfrak{P}}) \\ &= -\sum_{\mathfrak{Q}|\infty} \log |u_{\mathfrak{P}}^{\varepsilon_{\mathfrak{P}}-1}|_{\mathfrak{Q}}\mathfrak{Q} + (mh \log N(\mathfrak{P})\varepsilon_{\mathfrak{P}} + 1 - \varepsilon_{\mathfrak{P}})d_{\mathfrak{P}} \\ &\quad -(m\varepsilon_{\mathfrak{P}} + 1 - \varepsilon_{\mathfrak{P}}) \sum_{\mathfrak{Q}|\infty} \log |u_{\mathfrak{P}}|_{\mathfrak{Q}}\mathfrak{Q} \\ &= (\tilde{h} \log N(\mathfrak{P})\varepsilon_{\mathfrak{P}} + 1 - \varepsilon_{\mathfrak{P}})d_{\mathfrak{P}} - m\varepsilon_{\mathfrak{P}} \sum_{\mathfrak{Q}|\infty} \log |u_{\mathfrak{P}}|_{\mathfrak{Q}}\mathfrak{Q} \\ &= \tilde{\lambda}_S^{\mathrm{mod}}(1_{\mathfrak{P}}). \end{aligned}$$

We get

$$\begin{aligned} \frac{\det(\tilde{\lambda}_S^{\mathrm{mod}}\phi|\mathrm{Hom}_G(V_{\check{\chi}}, \mathbb{C}\overline{\nabla}_S))}{\det(\lambda_S^{\mathrm{mod}}\phi|\mathrm{Hom}_G(V_{\check{\chi}}, \mathbb{C}\overline{\nabla}_S))} &= \det(\psi|\mathrm{Hom}_G(V_{\check{\chi}}, \mathbb{C}(E_S \oplus C))) \\ &= \prod_{\mathfrak{P} \in S_{\mathrm{ram}}^* \setminus (S \cap S_{\mathrm{ram}})^*} m^{\dim V_{\check{\chi}}^{G_{\mathfrak{P}}}} \end{aligned}$$

as desired. □

The properties of the homomorphism A_ϕ^{mod} are summarized in the following

Theorem 1.5.8 *Fix a set of data (D). Let F be a number field, Galois over $\mathbb{Q}$ with Galois group Γ, which is large enough such that all representations of G can be realized over F. Then the following holds:*

(1) $A_\phi^{\mathrm{mod}}(\chi^\sigma) = A_\phi^{\mathrm{mod}}(\chi)^\sigma$ for all $\sigma \in \Gamma$ if and only if Stark's conjecture (1.5.1) holds.

(2) If χ is an irreducible symplectic character of G, then $A_\phi^{\mathrm{mod}}(\chi)W(\chi) \in \mathbb{R}^+$.

(3) If $\phi' : \mathbb{Q}\overline{\nabla} \to \mathbb{Q}(E_S \oplus C)$ is another $\mathbb{Q}G$-isomorphism, then

$$\frac{A^{\mathrm{mod}}_{\phi'}(\chi)}{A^{\mathrm{mod}}_{\phi}(\chi)} \equiv \det(\phi^{-1}\phi' | \mathrm{Hom}_G(V_{\check{\chi}}, \mathbb{C}\overline{\nabla})) \bmod \mathrm{Det}\,(U(\mathbb{Z}G)).$$

(4) Let $\mathfrak{P}_0$ be a prime not in S which ramifies in L/K. Given an integral multiple h of h_L, the class number of L, and $\mathbb{Q}G$-isomorphisms ϕ and ϕ_0 as in diagram (1.38) we have an equality

$$\frac{A^{\mathrm{mod}}_{\phi_0}(\chi)}{A^{\mathrm{mod}}_{\phi}(\chi)} \equiv (-h|G_{\mathfrak{P}_0}|)^{\dim V_{\check{\chi}}^{G_{\mathfrak{P}_0}}} \bmod \mathrm{Det}\,(U(\mathbb{Z}G)).$$

(5) Let $\mathfrak{P}_0$ be a prime not in S which does not ramify in L/K. Given $\mathbb{Q}G$-isomorphisms ϕ and ϕ_0 as in diagram (1.45) we have an equality

$$\frac{A^{\mathrm{mod}}_{\phi_0}(\chi)}{A^{\mathrm{mod}}_{\phi}(\chi)} \equiv (|G_{\mathfrak{P}_0}|)^{\dim V_{\check{\chi}}^{G_{\mathfrak{P}_0}}} \cdot \det(\phi_{\mathfrak{P}_0} - 1 | V_{\check{\chi}} / V_{\check{\chi}}^{G_{\mathfrak{P}_0}})^{-1} \bmod \mathrm{Det}\,(U(\mathbb{Z}G)).$$

Before proving the theorem, we now point out how to state the LRNC for small sets of places.
Assume that Stark's conjecture holds. By (1), (2) and Proposition 1.5.7 we can view the map

$$\chi \mapsto A^{\mathrm{mod}}_{\phi}(\check{\chi}) W(L/K, \check{\chi})$$

as a representing homomorphism of an element in $K_0T(\mathbb{Z}G)$ via the isomorphism (1.9). Since Theorem 1.5.8 together with Proposition 1.4.2, Theorem 1.4.3 and Theorem 1.4.4 show that this homomorphism exactly behaves like Ω_ϕ, it is now evident to state the

Conjecture 1.5.9 (LRNC for small S) *The element $\Omega_\phi \in K_0T(\mathbb{Z}G)$ has representing homomorphism $\chi \mapsto A^{\mathrm{mod}}_{\phi}(\check{\chi}) W(L/K, \check{\chi})$.*

Theorem 1.5.8 now implies the

Corollary 1.5.10 *The Lifted Root Number Conjecture for small sets of places is equivalent to the Lifted Root Number Conjecture for large sets of places.*

For this reason we refer to conjecture 1.5.9 as well as to conjecture 1.5.3 as the Lifted Root Number Conjecture.

The element Ω_ϕ decomposes into p-parts $\Omega_\phi^{(p)}$ via the isomorphism (1.7). If we choose a prime $\wp$ in F above p and an embedding $j_p : F \rightarrowtail F_\wp$ for each p, Stark's conjecture asserts that the map

$$(A^{\mathrm{mod}}_{\phi})^{(p)} : \chi \mapsto j_p(A^{\mathrm{mod}}_{\phi}(j_p^{-1}(\chi)))$$

lies in $\mathrm{Hom}_{\Gamma_\wp}(R_p(G), F_\wp^\times)$. Conjecture 1.5.9 localizes to

Conjecture 1.5.11 (LRNC for small S at the prime p) *The element $\Omega_\phi^{(p)} \in K_0T(\mathbb{Z}_pG)$ has representing homomorphism $\chi \mapsto (A_\phi^{\mathrm{mod}})^{(p)}(\check{\chi})$.*

We obviously have the

Corollary 1.5.12 *The Lifted Root Number Conjecture is true for L/K if and only if Conjecture 1.5.11 is true for L/K and all primes p.*

We conclude this section with the

Proof of Theorem 1.5.8. Because of the commutative triangle

$$\begin{array}{ccc} & & \mathbb{Q}\overline{\nabla} \\ & \swarrow^{\phi^{-1}\phi'} & \downarrow{\scriptstyle \phi'} \\ \mathbb{Q}\overline{\nabla} & \xrightarrow{\phi} & \mathbb{Q}\overline{\nabla} \end{array}$$

assertion (3) is clear, and since the map

$$\chi \mapsto \det(\phi^{-1}\phi' | \mathrm{Hom}_G(V_{\check{\chi}}, \mathbb{C}\overline{\nabla}))$$

commutes with the action of Γ, (1) is independent of the choice of ϕ. Hence, we may take an arbitrary embedding $\phi_S : \Delta S \rightarrowtail E_S$ and choose $\phi = \phi_\nabla$ fitting in a diagram

$$\begin{array}{ccc} \Delta_{\mathbb{Q}} S & \xrightarrow{\phi_S} & \mathbb{Q}E_S \\ \downarrow & & \downarrow \\ \mathbb{Q}\overline{\nabla} & \xrightarrow{\phi_\nabla} & \mathbb{Q}(E_S \oplus C) \\ \downarrow & & \downarrow \\ \bigoplus_{\mathfrak{P} \in S_{\mathrm{ram}}^* \setminus (S \cap S_{\mathrm{ram}})^*} \mathrm{ind}_{G_\mathfrak{P}}^G \mathbb{Q}W_\mathfrak{P}^0 & \xrightarrow{\phi_C} & \mathbb{Q}C \end{array}$$

where ϕ_C sends $1 \otimes d_\mathfrak{P}$ to $1_\mathfrak{P}$. After tensoring with $\mathbb{C}$ we can extend the above diagram to

$$\begin{array}{ccccc} \Delta_{\mathbb{C}} S & \xrightarrow{\phi_S} & \mathbb{C}E_S & \xrightarrow{\lambda_S} & \Delta_{\mathbb{C}} S \\ \downarrow & & \downarrow & & \downarrow \\ \mathbb{C}\overline{\nabla} & \xrightarrow{\phi_\nabla} & \mathbb{C}(E_S \oplus C) & \xrightarrow{\lambda_S^{\mathrm{mod}}} & \mathbb{C}\overline{\nabla} \\ \downarrow & & \downarrow & & \downarrow \\ \bigoplus_{\substack{\mathfrak{P} \in S_{\mathrm{ram}}^* \\ \mathfrak{P} \notin S^*}} \mathrm{ind}_{G_\mathfrak{P}}^G \mathbb{C}W_\mathfrak{P}^0 & \xrightarrow{\phi_C} & \mathbb{C}C & \xrightarrow{\overline{\lambda}_C} & \bigoplus_{\substack{\mathfrak{P} \in S_{\mathrm{ram}}^* \\ \mathfrak{P} \notin S^*}} \mathrm{ind}_{G_\mathfrak{P}}^G \mathbb{C}W_\mathfrak{P}^0 \end{array}$$

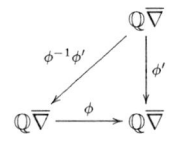

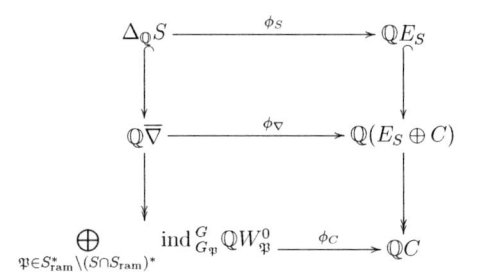

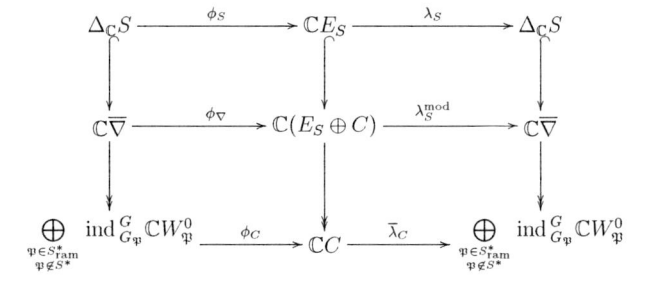

where $\overline{\lambda}_C$ is the map λ_C above composed with the canonical projection onto $\bigoplus \mathrm{ind}\,_{G_{\mathfrak{P}}}^{G} \mathbb{C}W_{\mathfrak{P}}^0$, hence

$$\overline{\lambda}_C(1_{\mathfrak{P}}) = \left(h \log N(\mathfrak{P}) \frac{1}{|G_{\mathfrak{P}}|} N_{G_{\mathfrak{P}}} + 1 - \frac{1}{|G_{\mathfrak{P}}|} N_{G_{\mathfrak{P}}} \right) d_{\mathfrak{P}}.$$

Thus, we get

$$\begin{aligned}
\frac{R_{\phi_\nabla}^{\mathrm{mod}}(\chi)}{R_{\phi_S}(\chi)} &= \frac{\det(\overline{\lambda}_C \phi_C | \mathrm{Hom}_G(V_{\check{\chi}}, \bigoplus \mathrm{ind}\,_{G_{\mathfrak{P}}}^{G} \mathbb{C}W_{\mathfrak{P}}^0))}{\prod_{\mathfrak{P} \in S_{\mathrm{ram}}^* \backslash (S \cap S_{\mathrm{ram}})^*} (-h|G_{\mathfrak{P}}|)^{\dim V_{\check{\chi}}^{G_{\mathfrak{P}}}}} \\
&= \prod_{\mathfrak{P} \in S_{\mathrm{ram}}^* \backslash (S \cap S_{\mathrm{ram}})^*} \frac{\det(h \log N(\mathfrak{P}) \frac{1}{|G_{\mathfrak{P}}|} N_{G_{\mathfrak{P}}} + 1 - \frac{1}{|G_{\mathfrak{P}}|} N_{G_{\mathfrak{P}}} | V_{\check{\chi}})}{(-h|G_{\mathfrak{P}}|)^{\dim V_{\check{\chi}}^{G_{\mathfrak{P}}}}} \\
&= \prod_{\mathfrak{P} \in S_{\mathrm{ram}}^* \backslash (S \cap S_{\mathrm{ram}})^*} \left(\frac{-\log N(\mathfrak{P})}{|G_{\mathfrak{P}}|} \right)^{\dim V_{\check{\chi}}^{G_{\mathfrak{P}}}} .
\end{aligned}$$

By Proposition 6 in [We], p. 50 we have

$$\frac{c_{S \cup S_{\mathrm{ram}}}(\chi)}{c_S(\chi)} = \prod_{\mathfrak{P} \in S_{\mathrm{ram}}^* \backslash (S \cap S_{\mathrm{ram}})^*} \log N(\mathfrak{p})^{\dim V_{\check{\chi}}^{G_{\mathfrak{P}}}} \det(1 - \phi_{\mathfrak{P}} | \dim V_{\check{\chi}}^{I_{\mathfrak{P}}} / V_{\check{\chi}}^{G_{\mathfrak{P}}}),$$

where $\mathfrak{p}$ is the prime in K below $\mathfrak{P}$. Writing $e_{\mathfrak{P}/\mathfrak{p}}$ for the ramification index of $\mathfrak{P}$ over $\mathfrak{p}$, we end up with

$$\frac{A_{\phi_S}(\chi)}{A_{\phi_\nabla}^{\mathrm{mod}}(\chi)} = \prod_{\mathfrak{P} \in S_{\mathrm{ram}}^* \backslash (S \cap S_{\mathrm{ram}})^*} (-e_{\mathfrak{P}/\mathfrak{p}})^{\dim V_{\check{\chi}}^{G_{\mathfrak{P}}}} \det(1 - \phi_{\mathfrak{P}} | \dim V_{\check{\chi}}^{I_{\mathfrak{P}}} / V_{\check{\chi}}^{G_{\mathfrak{P}}}).$$

Since the right hand side commutes with the action of Γ this completes the proof of (1).

If χ is an irreducible symplectic character one knows that $W(\chi)/c_S(\chi) \in \mathbb{R}^+$ for any set S and likewise $R_{\phi_S}(\chi) \in \mathbb{R}^+$ (cf. [We], Lemma 11c, p.50 and Proposition 7b, p. 57 resp. its proof). Since $\dim V_{\check{\chi}}^{G_{\mathfrak{P}}}$ is even in this case, we get (2). For (4) we consider the diagram

$$\begin{array}{ccccc}
\Delta G_{\mathfrak{P}_0} \cdot \mathbb{R}G & \hookrightarrow & \mathbb{R}(E_S \oplus C) & \xrightarrow{\delta} & \mathbb{R}(E_{S_0} \oplus C_0) \\
\phi'^{-1} \downarrow \simeq & & \downarrow \lambda_S^{\mathrm{mod}} & & \downarrow \lambda_{S_0}^{\mathrm{mod}} \\
\mathbb{R}G/N_{G_{\mathfrak{P}_0}} & \hookrightarrow & \mathbb{R}\overline{\nabla}_S & \xrightarrow{\pi_{\overline{\nabla}}} & \mathbb{R}\overline{\nabla}_{S_0}
\end{array}$$

where the upper sequence derives from (1.36) and the lower sequence from (1.35). The isomorphism ϕ' has been defined in (1.37). We have to check commutativity.
For the right hand square it suffices to show that

$$\lambda_{S_0}^{\mathrm{mod}}(\delta(1_{\mathfrak{P}_0})) = \pi_{\overline{\nabla}}(\lambda_S^{\mathrm{mod}}(1_{\mathfrak{P}_0})).$$

The projection $\pi_{\overline{\nabla}}$ is induced from $W^0_{\mathfrak{P}_0} \twoheadrightarrow \mathbb{Z}$ which maps $d_{\mathfrak{P}_0}$ to 1. Hence

$$\begin{aligned}
\pi_{\overline{\nabla}}(\lambda_S^{\mathrm{mod}}(1_{\mathfrak{P}_0})) &= \pi_{\overline{\nabla}}\left(\left(h \log N(\mathfrak{P}_0)\frac{1}{|G_{\mathfrak{P}_0}|}N_{G_{\mathfrak{P}_0}} + 1 - \frac{1}{|G_{\mathfrak{P}_0}|}N_{G_{\mathfrak{P}_0}}\right) d_{\mathfrak{P}_0}\right.\\
&\quad \left. - \sum_{\mathfrak{Q}|\infty} \log |u_{\mathfrak{P}_0}|_{\mathfrak{Q}}\mathfrak{Q}\right)\\
&= h \log N(\mathfrak{P}_0)\mathfrak{P}_0 - \sum_{\mathfrak{Q}|\infty} \log |u_{\mathfrak{P}_0}|_{\mathfrak{Q}}\mathfrak{Q}\\
&= \lambda_{S_0}^{\mathrm{mod}}(u_{\mathfrak{P}_0}, 0)\\
&= \lambda_{S_0}^{\mathrm{mod}}(\delta(1_{\mathfrak{P}_0}))
\end{aligned}$$

as desired. For the left hand square let $\alpha \in \Delta G_{\mathfrak{P}_0}$ and $x \in \mathbb{R}G$. We have to verify that $\lambda_S^{\mathrm{mod}}(u_{\mathfrak{P}_0}^{-\alpha x}, \alpha x \cdot 1_{\mathfrak{P}_0}) = \alpha x \cdot \kappa(1,0)$, where κ is the epimorphism from Proposition 1.5.4 for the prime $\mathfrak{P}_0$. But this is true, since

$$\begin{aligned}
\lambda_S^{\mathrm{mod}}(u_{\mathfrak{P}_0}^{-\alpha x}, \alpha x \cdot 1_{\mathfrak{P}_0}) &= -\sum_{\mathfrak{Q}|\infty} \log |u_{\mathfrak{P}_0}^{-\alpha x}|_{\mathfrak{Q}}\mathfrak{Q} + \alpha x(d_{\mathfrak{P}_0} - \sum_{\mathfrak{Q}|\infty} \log |u_{\mathfrak{P}_0}|_{\mathfrak{Q}}\mathfrak{Q})\\
&= \alpha x d_{\mathfrak{P}_0}\\
&= \alpha x \left(1 - \frac{1}{|G_{\mathfrak{P}_0}|}N_{G_{\mathfrak{P}_0}}\right) d_{\mathfrak{P}_0}
\end{aligned}$$

and

$$\begin{aligned}
\kappa(1,0) &= \kappa\left(1, \frac{1}{|G_{\mathfrak{P}_0}|}N_{G_{\mathfrak{P}_0}}\right) - \kappa\left(0, \frac{1}{|G_{\mathfrak{P}_0}|}N_{G_{\mathfrak{P}_0}}\right)\\
&= d_{\mathfrak{P}_0} - \frac{1}{|G_{\mathfrak{P}_0}|}N_{G_{\mathfrak{P}_0}} h_{\mathfrak{P}_0}^{-1} N_{I_{\mathfrak{P}_0}} d_{\mathfrak{P}_0}\\
&= \left(1 - \frac{1}{|G_{\mathfrak{P}_0}|}N_{G_{\mathfrak{P}_0}}\right) d_{\mathfrak{P}_0},
\end{aligned}$$

where $h_{\mathfrak{P}_0}$ has been defined in the proof of Lemma 1.5.5.
Now we can glue the above diagram and diagram (1.38):

$$\begin{array}{ccccc}
\mathbb{C}G/N_{G_{\mathfrak{P}_0}} & \xrightarrow{\phi'} & \Delta G_{\mathfrak{P}_0} \cdot \mathbb{C}G & \xrightarrow{\phi'^{-1}} & \mathbb{C}G/N_{G_{\mathfrak{P}_0}}\\
\downarrow & & \downarrow & & \downarrow\\
\mathbb{C}\overline{\nabla}_S & \xrightarrow{\phi} & \mathbb{C}(E_S \oplus C) & \xrightarrow{\lambda_S^{\mathrm{mod}}} & \mathbb{C}\overline{\nabla}_S\\
\downarrow & & \downarrow & & \downarrow\\
\mathbb{C}\overline{\nabla}_{S_0} & \xrightarrow{\phi_0} & \mathbb{C}(E_{S_0} \oplus C_0) & \xrightarrow{\lambda_{S_0}^{\mathrm{mod}}} & \mathbb{C}\overline{\nabla}_{S_0}
\end{array}$$

Thus, we get

$$\det(\lambda_S^{\mathrm{mod}}\phi|\mathrm{Hom}_G(V_{\check{\chi}}, \mathbb{C}\overline{\nabla}_S)) = \det(\lambda_{S_0}^{\mathrm{mod}}\phi_0|\mathrm{Hom}_G(V_{\check{\chi}}, \mathbb{C}\overline{\nabla}_{S_0}))$$

and $c_{S \cup S_{\mathrm{ram}}}(\chi) = c_{S_0 \cup S_{\mathrm{ram}}}(\chi)$, since $\mathfrak{P}_0$ ramifies in L/K. Now (4) is clear. For (5) we may assume that S contains all the ramified primes. Hence, by Proposition 5 in [GRW] or Proposition 8(b), p.11 in [We] (but observe that the Dirichlet map there is the negative of ours) we get

$$\frac{A_{\phi_0}^{\mathrm{mod}}(\chi)}{A_{\phi}^{\mathrm{mod}}(\chi)} = (|G_{\mathfrak{P}_0}|)^{\dim V_\chi^{G_{\mathfrak{P}_0}}} \cdot \det(1 - \phi_{\mathfrak{P}_0}|V_\chi/V_\chi^{G_{\mathfrak{P}_0}})^{-1}.$$

Furthermore, $\dim V_\chi^{G_{\mathfrak{P}_0}} = \dim V_{\check{\chi}}^{G_{\mathfrak{P}_0}}$ and

$$[\chi \mapsto \det(\phi_{\mathfrak{P}_0} - 1|V_{\check{\chi}}/V_{\check{\chi}}^{G_{\mathfrak{P}_0}}) \det(1 - \phi_{\mathfrak{P}_0}|V_\chi/V_\chi^{G_{\mathfrak{P}_0}})^{-1}] = [\chi \mapsto \det(\phi_{\mathfrak{P}}^{-1}|V_\chi)],$$

which lies in Det $(U(\mathbb{Z}G))$. This completes the proof of (5) and the theorem. □

In the next chapter we will give an application of conjecture 1.5.9 in the context of tame CM-extensions.

Chapter 2

Tame CM-extensions

In this chapter we apply the results of the previous section to CM-extensions of number fields which will soon assumed to be tame above a fixed rational prime $p \neq 2$.
So let L/K be a CM-extension, i.e. K is totally real and L is a totally imaginary quadratic extension of a totally real number field. Complex conjugation on $\mathbb{C}$ induces an automorphism on L which is independent of the embedding into $\mathbb{C}$ (cf. [Wa], p. 38). We denote this automorphism by j and refer to it as complex conjugation as well. If L/K is Galois with Galois group G, this automorphism lies in the center of G.

For any G-module M we define submodules

$$M^+ := \{m \in M : jm = m\},$$

$$M^- := \{m \in M : jm = -m\}.$$

M^+ is a module over the ring $\mathbb{Z}G_+ := \mathbb{Z}G/(1-j) = \mathbb{Z}[G/\langle j \rangle]$, whereas M^- has a $\mathbb{Z}G_- := \mathbb{Z}G/(1+j)$ action, but $\mathbb{Z}G_-$ is not a ring, since $\frac{1-j}{2} \notin \mathbb{Z}G_-$.

Examples.

(1) For $M = \mathbb{Z}G$ we have $\mathbb{Z}G^- = (1-j)\mathbb{Z}G$ and multiplication by $(1-j)$ induces an isomorphism

$$\mathbb{Z}G_- \simeq \mathbb{Z}G^-.$$

(2) If we apply the $+$ functor to $M = L$, we get the uniquely determined maximal real subfield L^+ of L.

(3) If $M = \mathfrak{o}_L^\times$, the global units of L, the minus part of M is just the kernel of the Dirichlet map, which consists of the roots of unity in L. We denote these by μ_L and thus

$$(\mathfrak{o}_L^\times)^- = \mu_L.$$

(4) Let us denote the set of all infinite primes of L by S_∞. Since j acts on S_∞ as the identity, we have

$$(\Delta S_\infty)^- = (\mathbb{Z}S_\infty)^- = 0.$$

(5) If M is any $\langle j\rangle$-module and $M(1)$ is the twisted $\langle j\rangle$-module, i.e. $M = M(1)$ as sets and j acts on $M(1)$ such as $-j$ on M, we have

$$M^- = M(1)^+.$$

Let R be a number field or (a localization of) the ring of integers of a number field. An exact sequence $A \rightarrowtail B \twoheadrightarrow C$ of RG-modules gives rise to a long exact sequence

$$\begin{aligned} A^- \rightarrowtail B^- \to C^- &\to H^0(\langle j\rangle, A) \to H^0(\langle j\rangle, B) \to H^0(\langle j\rangle, C) \\ &\to H^1(\langle j\rangle, A) \to \cdots, \end{aligned}$$

where we make the convention that all occurring cohomology groups are Tate cohomology groups if not otherwise stated. Indeed, by example (5) we get a long exact sequence

$$\begin{aligned} A^- \rightarrowtail B^- \to C^- &\to H^1(\langle j\rangle, A(1)) \to H^1(\langle j\rangle, B(1)) \to H^1(\langle j\rangle, C(1)) \\ &\to H^2(\langle j\rangle, A(1)) \to \cdots \end{aligned}$$

and for any G-module M and for all $i \in \mathbb{Z}$ we have isomorphisms

$$H^i(\langle j\rangle, M) \simeq H^{i+1}(\langle j\rangle, M(1)).$$

Since $\langle j\rangle$ is cyclic and $M(1)(1) = M$ it suffices to check this for $i = -1$, and in fact

$$H^{-1}(\langle j\rangle, M) = M^-/(1-j)M = M(1)^+/(1+j)M(1) = H^0(\langle j\rangle, M(1)).$$

Hence, the minus functor is left exact, and even exact if 2 is invertible in R.

If a finitely generated G-module M decomposes in

$$M = M^+ \oplus M^-,$$

the natural maps

$$\begin{aligned} H^i(U, M^+) &\to H^i(U, M)^+, \\ H^i(U, M^-) &\to H^i(U, M)^- \end{aligned}$$

are isomorphisms for all subgroups U of G of odd order, $i \in \mathbb{Z}$. Indeed, the composite map

$$H^i(U, M) \simeq H^i(U, M^+) \oplus H^i(U, M^-) \to H^i(U, M)^+ \oplus H^i(U, M)^- \simeq H^i(U, M)$$

is the identity, Here, the rightmost isomorphism exists, because $H^i(U, M)$ is finite of odd order and hence also decomposes in a plus and a minus part.

If $p \neq 2$ and M is a $\mathbb{Z}_pG$-module, there is a natural decomposition

$$M = M^+ \oplus M^-$$

which induces an isomorphism

$$K_0T(\mathbb{Z}_pG) \simeq K_0T(\mathbb{Z}_pG_+) \oplus K_0T(\mathbb{Z}_pG_-). \tag{2.1}$$

These isomorphisms combine to an isomorphism

$$K_0T(\mathbb{Z}[\tfrac{1}{2}]G) \simeq K_0T(\mathbb{Z}[\tfrac{1}{2}]G_+) \oplus K_0T(\mathbb{Z}[\tfrac{1}{2}]G_-).$$

We recall some notation to describe the isomorphism (2.1) in terms of representing homomorphisms. Let F be a number field which is large enough such that all representations of G can be realized over F and which is Galois over $\mathbb{Q}$ with Galois group Γ. Choose a prime $\wp$ in F above p and denote the ring of virtual characters of G with values in $\mathbb{Q}_p^c$ by $R_p(G)$. By (1.8) the elements in $K_0T(\mathbb{Z}_pG)$ are represented by homomorphisms in $\mathrm{Hom}_{\Gamma_\wp}(R_p(G), F_\wp^\times)$.
A character χ is called even if $\chi(j) = \chi(1)$, and it is called odd if $\chi(j) = -\chi(1)$.
Let us define $R_p^+(G)$ and $R_p^-(G)$ to be the subrings of $R_p(G)$ generated by even and odd characters, respectively. The Hom description and the above isomorphism now give

$$\frac{\mathrm{Hom}_{\Gamma_\wp}(R_p(G), F_\wp^\times)}{\mathrm{Det}\,(\mathbb{Z}_pG^\times)} \simeq \frac{\mathrm{Hom}_{\Gamma_\wp}(R_p^+(G), F_\wp^\times)}{\mathrm{Det}\,(\mathbb{Z}_pG_+^\times)} \oplus \frac{\mathrm{Hom}_{\Gamma_\wp}(R_p^-(G), F_\wp^\times)}{\mathrm{Det}\,(\mathbb{Z}_pG_-^\times)},$$

induced by the canonical restriction maps.

We denote the image of $\Omega_\phi^{(p)}$ in $K_0T(\mathbb{Z}_pG_+)$ and $K_0T(\mathbb{Z}_pG_-)$ by $\Omega_\phi^{(p),+}$ and $\Omega_\phi^{(p),-}$, respectively. Accordingly, the LRNC at p decomposes into a plus part and a minus part:

Proposition 2.0.13 *Let $p \neq 2$ be a rational prime and L/K a Galois CM-extension with Galois group G. The LRNC at p (Conjecture 1.5.11) is true if and only if the following two assertions hold*

(1) $\Omega_\phi^{(p),+}$ has representing homomorphism

$$[\chi \mapsto (A_\phi^{\mathrm{mod}})^{(p)}(\check{\chi})] \in \mathrm{Hom}_{\Gamma_\wp}(R_p^+(G), F_\wp^\times).$$

(2) $\Omega_\phi^{(p),-}$ has representing homomorphism

$$[\chi \mapsto (A_\phi^{\mathrm{mod}})^{(p)}(\check{\chi})] \in \mathrm{Hom}_{\Gamma_\wp}(R_p^-(G), F_\wp^\times).$$

In the following, we only deal with the minus part of the LRNC.

For later use we state the following Lemma, which is taken from [Ch2], p.369.

Lemma 2.0.14 *Let L/K be a tame Galois extension of number fields and $\mathfrak{P}$ a finite prime of L. Then the inertia group $I_{\mathfrak{P}}$ is cyclic and we choose a generator a of $I_{\mathfrak{P}}$. Let $b \in G_{\mathfrak{P}}$ be a lift of the automorphism $\phi_{\mathfrak{P}}^{-1} \in G_{\mathfrak{P}}/I_{\mathfrak{P}}$ which is of maximal order among all such elements. Define $e_{\mathfrak{P}} = |I_{\mathfrak{P}}|$, $f_{\mathfrak{P}} = |G_{\mathfrak{P}}/I_{\mathfrak{P}}|$ and $q_{\mathfrak{p}} = |\mathfrak{o}_K/\mathfrak{p}|$, where $\mathfrak{p} = \mathfrak{P} \cap K$.*
Then $G_{\mathfrak{P}}$ is generated by a and b, and

$$\begin{aligned} ab &= ba^{q_{\mathfrak{p}}} \\ b^{f_{\mathfrak{P}}} &= a^{c_{\mathfrak{P}}} \end{aligned}$$

for some integer $c_{\mathfrak{P}} \mid e_{\mathfrak{P}}$.

2.1 Ray class groups

Let L/K be a Galois CM-extension with Galois group G. The class group cl_L occurs in the construction of a Tate-sequence for S_∞, as it is the torsion submodule of ∇. Hence, one expects a relation between the LRNC and cl_L. But cl_L rarely is c.t.; so we intend to replace it by an appropriate c.t. ray class group.

If T is a finite G-invariant set of non-archimedean places of L we write cl_L^T for the ray class group to the ray $\mathfrak{M}_T := \prod_{\mathfrak{P}\in T} \mathfrak{P}$. Let S be a second finite G-invariant set of places of L which contains all the archimedean primes and satisfies $S \cap T = \emptyset$. We write S_f for the set of all finite primes in S. There is a natural map $\mathbb{Z}S_f \to \mathrm{cl}_L^T$ which sends each prime $\mathfrak{P} \in S_f$ to the corresponding class $[\mathfrak{P}] \in \mathrm{cl}_L^T$. We denote the cokernel of this map by cl_S^T. Further, define

$$E_S^T := \{x \in E_S : x \equiv 1 \bmod \mathfrak{M}_T\}.$$

Since the sets S and T are both G-invariant, all these modules are equipped with a natural G-action. Hence, we have the following exact sequences of G-modules

$$E_{S_\infty}^T \rightarrowtail E_S^T \xrightarrow{v} \mathbb{Z}S_f \to \mathrm{cl}_L^T \twoheadrightarrow \mathrm{cl}_S^T, \tag{2.2}$$

where $v(x) = \sum_{\mathfrak{P}\in S_f} v_{\mathfrak{P}}(x)\mathfrak{P}$ for $x \in E_S^T$, and

$$E_S^T \rightarrowtail E_S \to (\mathfrak{o}_S/\mathfrak{M}_T)^\times \xrightarrow{\nu} \mathrm{cl}_S^T \twoheadrightarrow \mathrm{cl}_S, \tag{2.3}$$

where the map ν lifts an element $\overline{x} \in (\mathfrak{o}_S/\mathfrak{M}_T)^\times$ to $x \in \mathfrak{o}_S$ and sends it to the ideal class $[(x)] \in \mathrm{cl}_S^T$ of the principal ideal (x). We define

$$A_S^T := (\mathrm{cl}_S^T)^-.$$

If $S = S_\infty$, we also write A_L^T and E_L^T instead of $A_{S_\infty}^T$ and $E_{S_\infty}^T$.

Since $(\mathfrak{o}_L^\times)^- = \mu_L$, one can always find primes $\mathfrak{P}$ of L such that $(E_L^T)^- = 1$ for all sets of places T with $\mathfrak{P} \in T$. One only has to check if $1-\zeta \notin \prod_{g\in G/G_\mathfrak{P}} \mathfrak{P}^g$ for all $\zeta \in \mu_L$, $\zeta \neq 1$; this is true for all but finitely many primes of L.

The main result of this section is

Theorem 2.1.1 *Let L/K be a Galois CM-extension with Galois group G, $p \neq 2$ a rational prime and $S_p = \{\mathfrak{P} \subset L : \mathfrak{P} \mid p\}$. Assume that for all $\mathfrak{P} \in S_p \cap S_{\text{ram}}$ the ramification is tame or $j \in G_\mathfrak{P}$. Choose a prime $\mathfrak{P}_0$ of L such that $1 - \zeta \notin \prod_{g\in G/G_{\mathfrak{P}_0}} \mathfrak{P}_0^g$ for all $\zeta \in \mu_L$, $\zeta \neq 1$.*
Then $A_L^T \otimes \mathbb{Z}_p$ is a c.t. G-module for each finite G-invariant set T of places of L that contains $\mathfrak{P}_0$ and all the ramified primes which are not in S_p.

REMARK. If L/K is tame above p and G is abelian, the above theorem follows from the proof of Proposition 7 in [Gr2]. The condition $j \in G_\mathfrak{P}$ is technical; but it sometimes is useful that j acts on local objects. The following proof is a good example.

PROOF. It suffices to show that $H^i(P, A_L^T \otimes \mathbb{Z}_p) = 1$ for $i \in \mathbb{Z}$ and all q-Sylow subgroups P of G. This is clear for $q \neq p$. So let P be a p-Sylow subgroup.
For any prime $\mathfrak{P}$ of L we write $U_\mathfrak{P}^0$ for the group of local units of the completion $L_\mathfrak{P}$ of L at $\mathfrak{P}$. Furthermore, we denote the group of local units congruent to 1 mod $\mathfrak{P}^n$ by $U_\mathfrak{P}^n$. Let us define an idèle subgroup

$$J_L^T := \prod_{\mathfrak{P}\in T} U_\mathfrak{P}^1 \times \prod_{\mathfrak{P}\notin T} U_\mathfrak{P}^0.$$

The following exact sequences define C_L^T:

$$E_L^T \rightarrowtail J_L^T \twoheadrightarrow C_L^T, \tag{2.4}$$

$$C_L^T \rightarrowtail C_L \twoheadrightarrow \mathrm{cl}_L^T. \tag{2.5}$$

For both sequences we take the long exact sequence in homology with respect to P. Thereafter, we apply the minus functor, which is exact in this case, since all the occurring homology groups are finite of odd order. The fact that $\mathfrak{P}_0 \in T$ forces

$$H^i(P, E_L^T)^- = H^i(P, (E_L^T)^-) = H^i(P, 1) = 1,$$

and hence sequence (2.4) implies

$$H^i(P, J_L^T)^- \simeq H^i(P, C_L^T)^-.$$

Global class field theory admits a similar argument for sequence (2.5):

$$H^i(P, C_L)^- \simeq H^{i-2}(P, \mathbb{Z})^- = H^{i-2}(P, \mathbb{Z}^-) = H^{i-2}(P, 0) = 1$$

and we therefore get isomorphisms

$$H^{i+1}(P, C_L^T)^- \simeq H^i(P, \mathrm{cl}_L^T)^- = H^i(P, \mathrm{cl}_L^T \otimes \mathbb{Z}_p)^- = H^i(P, A_L^T \otimes \mathbb{Z}_p).$$

Hence, it suffices to show that $H^i(P, J_L^T)^- = 1$ for all $i \in \mathbb{Z}$. The unit groups $U_{\mathfrak{P}}^n$ are c.t. $P_{\mathfrak{P}}$-modules if $\mathfrak{P}$ does not ramify in L/K. Even before taking minus parts, we thus get an isomorphism

$$H^i(P, J_L^T) \simeq \prod_{\mathfrak{p} \in S_{\mathrm{ram}}(K)} H^i(P, \prod_{\mathfrak{P}|\mathfrak{p}} U_{\mathfrak{P}}^{n_{\mathfrak{P}}}),$$

where $n_{\mathfrak{P}}$ is equal to 1 or 0 depending on wether $\mathfrak{P} \in T$ or not. If $\mathfrak{p}$ lies over a rational prime $q \neq p$, we have $n_{\mathfrak{P}} = 1$ for all $\mathfrak{P} \mid \mathfrak{p}$ by assumption. But in this case the unit groups $U_{\mathfrak{P}}^1$ are pro-q-groups and thus $H^i(P, \prod_{\mathfrak{P}|\mathfrak{p}} U_{\mathfrak{P}}^1) = 1$.
We are left with the case $\mathfrak{P} \in S_{\mathrm{ram}} \cap S_p$. For this, let F be the fixed field of P, and indicate the primes in F by a subscript F. We have

$$H^i(P, \prod_{\mathfrak{P}|\mathfrak{p}} U_{\mathfrak{P}}^{n_{\mathfrak{P}}}) \simeq \prod_{\mathfrak{p}_F|\mathfrak{p}} H^i(P, \prod_{\mathfrak{P}|\mathfrak{p}_F} U_{\mathfrak{P}}^{n_{\mathfrak{P}}}) = \prod_{\mathfrak{p}_F|\mathfrak{p}} H^i(P_{\mathfrak{P}}, U_{\mathfrak{P}}^{n_{\mathfrak{P}}}).$$

If $\mathfrak{P}$ is tamely ramified, it cannot ramify in L/F, since $P_{\mathfrak{P}}$ is a p-group. Hence, we get $H^i(P_{\mathfrak{P}}, U_{\mathfrak{P}}^{n_{\mathfrak{P}}}) = 1$ in this case. If otherwise $j \in G_{\mathfrak{P}}$, the action of j commutes with the above isomorphism, and we have to show that $H^i(P_{\mathfrak{P}}, U_{\mathfrak{P}}^{n_{\mathfrak{P}}})^- = 1$, $n_{\mathfrak{P}} \in \{0, 1\}$. By local class field theory

$$H^i(P_{\mathfrak{P}}, L_{\mathfrak{P}}^\times)^- \simeq H^{i-2}(P_{\mathfrak{P}}, \mathbb{Z})^- = H^{i-2}(P_{\mathfrak{P}}, \mathbb{Z}^-) = H^{i-2}(P_{\mathfrak{P}}, 0) = 1$$

and hence the short exact sequence

$$U_{\mathfrak{P}} \rightarrowtail L_{\mathfrak{P}}^\times \twoheadrightarrow \mathbb{Z}$$

implies $H^i(P_{\mathfrak{P}}, U_{\mathfrak{P}})^- = 1$. Finally, the sequence

$$U_{\mathfrak{P}}^1 \rightarrowtail U_{\mathfrak{P}} \twoheadrightarrow (\mathfrak{o}/\mathfrak{P})^\times$$

forces $H^i(P_{\mathfrak{P}}, U_{\mathfrak{P}}^1)^- = H^i(P_{\mathfrak{P}}, U_{\mathfrak{P}})^- = 1$, since the order of $(\mathfrak{o}/\mathfrak{P})^\times$ is relatively prime to p, and hence $H^i(P_{\mathfrak{P}}, (\mathfrak{o}/\mathfrak{P})^\times) = 1$. □

2.2 L-series and Stickelberger elements

In this section we fix, as before, a Galois CM-extension L/K of number fields with Galois group G and denote the complex conjugation on L by j. Let $w_L = |\mu_L|$ be the number of roots of unity in L and

$$Q := [\mathfrak{o}_L^\times : \mu_L \mathfrak{o}_{L^+}^\times] \in \{1, 2\}.$$

For the fact that Q equals 1 or 2 see [Wa], Theorem 4.12. By loc.cit. Theorem 4.10 the class number of L^+ divides the class number of L. The quotient h_L^- is called the relative class number.
For any finite set S of places of L and any character χ of G we denote the S-truncated L-function associated to χ by $L_S(L/K, \chi, s)$. Furthermore, the completed Artin L-series is defined to be

$$\Lambda(L/K, \chi, s) = c(L/K, \chi)^{s/2} \mathfrak{L}_\infty(L/K, \chi, s) L_{S_\infty}(L/K, \chi, s),$$

where

$$\begin{aligned} c(L/K, \chi) &= |d_K|^{\chi(1)} N(\mathfrak{f}(\chi)) \\ \mathfrak{L}_\infty(L/K, \chi, s) &= \begin{cases} L_{\mathbb{R}}(s)^{|S_\infty(K)|\chi(1)} & \text{if } \chi \text{ is even} \\ L_{\mathbb{R}}(s+1)^{|S_\infty(K)|\chi(1)} & \text{if } \chi \text{ is odd} \end{cases} \\ L_{\mathbb{R}}(s) &= \pi^{-s/2}\Gamma(s/2). \end{aligned}$$

Here, d_K is the discriminant of the number field K, $\mathfrak{f}(\chi)$ the Artin conductor of the character χ and $\Gamma(s)$ the usual complex Gamma function. The completed Artin L-series satisfies the functional equation

$$\Lambda(L/K, \chi, s) = W(\chi)\Lambda(L/K, \check{\chi}, 1 - s), \qquad (2.6)$$

where $W(\chi)$ is the Artin root number of the character χ and has absolute value 1 (cf. [Neu], Kap. VII, Theorem (12.6)).
Let $\operatorname{Irr}(G)$ be the set of irreducible characters of G and denote the trivial character by 1_G.

We now prove the following result:

Proposition 2.2.1 *Let L/K be a Galois CM-extension of number fields with Galois group G. Keeping the above notation we have*

$$\prod_{\substack{\chi \in \operatorname{Irr}(G) \\ \chi \text{ odd}}} L_{S_\infty}(L/K, \chi, 0)^{\chi(1)} = \pm \frac{2^{|S_\infty|} \cdot h_L^-}{Q \cdot w_L},$$

where the product runs through all the odd irreducible characters of G.

PROOF. Let us denote the Riemann zeta function of a number field F by $\zeta_F(s)$. We have (cf. [Neu], Kap. VII, Korollar (10.5))

$$\begin{aligned} \zeta_L(s) &= \zeta_K(s) \prod_{1_G \neq \chi \in \operatorname{Irr}(G)} L_{S_\infty}(L/K, \chi, s)^{\chi(1)} \\ \zeta_{L^+}(s) &= \zeta_K(s) \prod_{\substack{1_G \neq \chi \in \operatorname{Irr}(G) \\ \chi \text{ even}}} L_{S_\infty}(L/K, \chi, s)^{\chi(1)} \end{aligned}$$

Taking residuals at $s=1$ of both sides in these equations yields

$$\frac{(2\pi)^{|S_\infty|}\cdot h_L R_L}{w_L\sqrt{|d_L|}} = \operatorname{res}_{s=1}\zeta_K(s) \prod_{1_G\neq\chi\in\mathrm{Irr}\,(G)} L_{S_\infty}(L/K,\chi,1)^{\chi(1)}$$

$$\frac{2^{|S_\infty|}\cdot h_{L^+}R_{L^+}}{2\sqrt{|d_{L^+}|}} = \operatorname{res}_{s=1}\zeta_K(s) \prod_{\substack{1_G\neq\chi\in\mathrm{Irr}\,(G)\\ \chi\text{ even}}} L_{S_\infty}(L/K,\chi,1)^{\chi(1)},$$

where R_L and R_{L^+} are the regulators of L and L^+, respectively. If we divide the first by the second equation, we get by [Wa], Proposition 4.16

$$\frac{(2\pi)^{|S_\infty|}\cdot h_L^-}{Qw_L\sqrt{|d_L/d_{L^+}|}} = \prod_{\substack{\chi\in\mathrm{Irr}\,(G)\\ \chi\text{ odd}}} L_{S_\infty}(L/K,\chi,1)^{\chi(1)}.$$

Specializing the functional equation (2.6) at $s=1$ for odd characters χ,

$$c(L/K,\chi)^{1/2}\pi^{-|S_\infty(K)|\chi(1)}L_{S_\infty}(L/K,\chi,1) = W(\chi)L_{S_\infty}(L/K,\check{\chi},0),$$

gives

$$\begin{aligned}
\frac{(2\pi)^{|S_\infty|}\cdot h_L^-}{Qw_L\sqrt{|d_L/d_{L^+}|}} &= \prod_{\substack{\chi\in\mathrm{Irr}\,(G)\\ \chi\text{ odd}}} \left(L_{S_\infty}(L/K,\check{\chi},0)W(\chi)c(L/K,\chi)^{-1/2}\pi^{|S_\infty(K)|\chi(1)}\right)^{\chi(1)}\\
&\overset{(1)}{=} \frac{\pi^{|S_\infty|}}{\sqrt{|d_K|^{|G|/2}}} \prod_{\substack{\chi\in\mathrm{Irr}\,(G)\\ \chi\text{ odd}}} \left(L_{S_\infty}(L/K,\chi,0)W(\chi)N(\mathfrak{f}(\chi))^{-1/2}\right)^{\chi(1)}\\
&\overset{(2)}{=} \pm\frac{\pi^{|S_\infty|}}{\sqrt{|d_K|^{|G|/2}}} \prod_{\substack{\chi\in\mathrm{Irr}\,(G)\\ \chi\text{ odd}}} \left(L_{S_\infty}(L/K,\chi,0)N(\mathfrak{f}(\chi))^{-1/2}\right)^{\chi(1)}
\end{aligned}$$

Equality (1) holds, since $\sum_{\chi\text{ odd}}\chi(1)^2 = |G|/2$ and $|S_\infty(K)|\cdot|G|/2 = |S_\infty|$. As the product $\prod_{\chi\text{ odd}}W(\chi)$ is real and has absolute value 1, it equals ±1 and we get (2).

Let us write $\delta_{E/F}$ for the relative discriminant of an extension E/F of number fields, in particular $\delta_{E/\mathbb{Q}} = (d_E)$. We now compute

$$\begin{aligned}
\prod_{\substack{\chi\in\mathrm{Irr}\,(G)\\ \chi\text{ odd}}} N(\mathfrak{f}(\chi))^{\chi(1)} &= \frac{\prod_{\chi\in\mathrm{Irr}\,(G)} N(\mathfrak{f}(\chi))^{\chi(1)}}{\prod_{\substack{\chi\in\mathrm{Irr}\,(G)\\ \chi\text{ even}}} N(\mathfrak{f}(\chi))^{\chi(1)}} \overset{(1)}{=} \frac{N(\delta_{L/K})}{N(\delta_{L^+/K})}\\
&\overset{(2)}{=} N(\delta_{L^+/K})N(\delta_{L/L^+}) \overset{(2)}{=} N(\delta_{L^+/K})\frac{|d_L|}{|d_{L^+}|^2}\\
&\overset{(2)}{=} \frac{|d_L|}{|d_{L^+}|\cdot|d_K|^{|G|/2}}.
\end{aligned}$$

Equality (1) follows from the "Führerdiskriminantenproduktformel" (cf. [Neu], Kap. VII, (11.9)). For the equalities (2) note that in any tower $F \subset E \subset M$ of number fields we have $\delta_{M/F} = \delta_{E/F}^{[M:E]} N_{E/F}(\delta_{M/E})$.
If we put this in the previous equation, we obtain the desired result. □

For each irreducible character χ of G define

$$\varepsilon_\chi := \frac{\chi(1)}{|G|} \sum_{g \in G} \chi(g^{-1}) g.$$

The ε_χ are orthogonal central idempotents of $\mathbb{C}G$. Each generates one of the minimal ideals of the center of $\mathbb{C}G$, hence

$$Z(\mathbb{C}G) = \bigoplus_{\chi \in \mathrm{Irr}\,(G)} \mathbb{C}\varepsilon_\chi.$$

We define the following variant of a Stickelberger element which is closely related to the non-abelian Stickelberger-functions defined in [Ha]:

$$\omega := \sum_{\chi \in \mathrm{Irr}\,(G)} L_{S_\infty}(L/K, \check{\chi}, 0) \varepsilon_\chi \in Z(\mathbb{C}G) \tag{2.7}$$

Each $\mathbb{C}$-valued function on G extends to a $\mathbb{C}$-linear function on $\mathbb{C}G$. In particular, this applies to the irreducible characters of G, and obviously

$$\chi(\omega) = \chi(1) L_{S_\infty}(L/K, \check{\chi}, 0).$$

This property uniquely defines ω. If G is abelian, this element coincides with the element ω defined in [Gr3]. A priori, ω is an element of the group ring $\mathbb{C}G$, but we actually have

Proposition 2.2.2 *$\omega \in Z(\mathbb{Q}G)$, and even $\omega \in Z(\mathbb{Q}G^-)^\times$ if $|S_\infty| > 1$.*

PROOF. Note that the vanishing order of $L_{S_\infty}(L/K, \chi, s)$ in $s = 0$ equals

$$r_{S_\infty}(\chi) = \sum_{\mathfrak{P} \in S_\infty} \dim V_\chi^{G_\mathfrak{P}} - \dim V_\chi^G$$

by [Ta2], Proposition 3.4, p. 24. Hence, $L_{S_\infty}(L/K, \chi, 0) \neq 0$ if and only if χ is odd or χ is the trivial character and $|S_\infty| = 1$. This shows $\omega \in Z(\mathbb{C}G^-)^\times$ if $|S_\infty| > 1$. The coefficient of ω at $g \in G$ equals

$$\sum_{\chi \in \mathrm{Irr}\,(G)} L_{S_\infty}(L/K, \check{\chi}, 0) \frac{\chi(1)}{|G|} \chi(g^{-1})$$

which is invariant under Galois action, since $L_{S_\infty}(L/K, \check{\chi}, 0)^\sigma = L_{S_\infty}(L/K, \check{\chi}^\sigma, 0)$ for all $\sigma \in \mathrm{Gal}(\mathbb{Q}^c/\mathbb{Q})$ by Stark's conjecture, which is a theorem for odd characters and the trivial character (cf. [Ta2] Th. 1.2, p. 70 and Prop. 1.1, p. 44). □

Note that the proof also shows that in any case $\frac{1-j}{2}\omega \in Z(\mathbb{Q}G^-)^\times$.

Definition 2.2.3 *Let L/K be a Galois CM-extension with Galois group G and S, T be G-invariant sets of places of L. We define a Stickelberger element $\theta_S^T \in Z(\mathbb{C}G)$ by*

$$\chi(\theta_S^T) = \chi(\omega) \prod_{\mathfrak{P} \in T^*} \det(1 - \phi_{\mathfrak{P}}^{-1} q_{\mathfrak{p}} | V_\chi^{I_{\mathfrak{P}}}) \prod_{\mathfrak{P} \in S^*} \det(1 - \phi_{\mathfrak{P}}^{-1} | V_\chi^{I_{\mathfrak{P}}} / V_\chi^{G_{\mathfrak{P}}}),$$

where $\mathfrak{p} = \mathfrak{P} \cap K$ and $q_{\mathfrak{p}} = N(\mathfrak{p})$.

Since $\chi(\theta_S^T)$ differs from $\chi(\omega)$ by a factor which commutes with Galois action for each odd irreducible character χ, it follows from Proposition 2.2.2 that $\frac{1-j}{2}\theta_S^T \in Z(\mathbb{Q}G^-)^\times$. This enables us to make the following

Definition 2.2.4 *Let $F/\mathbb{Q}$ be a finite Galois extension with Galois group Γ such that each odd character of G can be realized over F. Then we define $\Theta_S^T \in \mathrm{Hom}_\Gamma(R^-(G), F^\times)$ by declaring*

$$\Theta_S^T(\chi) = \chi(1)^{-1}\chi(\theta_S^T)$$

on irreducible odd characters χ.

To afford an easier reading we will refer to the following setting as $(*)$:

- L/K is a Galois CM-extension with Galois group G.
- $p \neq 2$ is a rational prime.
- $S_p = \{\mathfrak{P} \subset L : \mathfrak{P} \mid p\}$
- Each $\mathfrak{P} \in S_p \cap S_{\mathrm{ram}}$ is at most tamely ramified or $j \in G_{\mathfrak{P}}$.
- $\mathfrak{P}_0$ is a prime of L, unramified in L/K such that $1 - \zeta \notin \prod_{g \in G/G_{\mathfrak{P}_0}} \mathfrak{P}_0^g$ for all $\zeta \in \mu_L$, $\zeta \neq 1$.
- T is a finite G-invariant set of places of L that contains $\mathfrak{P}_0$ and all the ramified primes which are not in S_p; $T \cap S_p = \emptyset$.
- S_1 is the set of all wildly ramified primes above p.

There is the following correspondence between the Stickelberger elements and the ray class groups $A_L^T \otimes \mathbb{Z}_p$ as defined in Theorem 2.1.1.

Proposition 2.2.5 *Fix a setting $(*)$. Then there exists an $\alpha \in \mathbb{Z}_p^\times$ such that*

$$|A_L^T \otimes \mathbb{Z}_p| = \alpha \cdot \prod_{\substack{\chi \in \mathrm{Irr}\,(G) \\ \chi \,\mathrm{odd}}} (\Theta_{S_1}^T(\chi))^{\chi(1)}.$$

Moreover, if G is abelian, we have $\frac{1-j}{2}\theta_{S_1}^T \in \mathbb{Z}_p G^-$ and

$$|A_L^T \otimes \mathbb{Z}_p| = |(\mathbb{Z}_p G)_- / \theta_{S_1}^T (\mathbb{Z}_p G)_-|.$$

PROOF. For an integer $m \in \mathbb{Z}$ let $m_p := p^{v_p(m)}$. Then the minus part of sequence (2.3) for $S = S_\infty$ tensored with $\mathbb{Q}$, namely

$$\mu_L \otimes \mathbb{Z}_p \rightarrowtail (\mathfrak{o}_L/\mathfrak{M}_T)^{\times,-} \otimes \mathbb{Z}_p \to A_L^T \otimes \mathbb{Z}_p \twoheadrightarrow \mathrm{cl}_L^- \otimes \mathbb{Z}_p, \tag{2.8}$$

implies the equality

$$|A_L^T \otimes \mathbb{Z}_p| = |A_L^T|_p = \frac{h_{L,p}^-}{w_{L,p}} |(\mathfrak{o}_L/\mathfrak{M}_T)^{\times,-}|_p. \tag{2.9}$$

Let us write $a \sim b$ if $ab^{-1} \in \mathbb{Z}_p^\times$. Then

$$\prod_{\substack{\chi \in \mathrm{Irr}\,(G) \\ \chi\,\mathrm{odd}}} (\chi(1)^{-1}\chi(\omega))^{\chi(1)} = \prod_{\substack{\chi \in \mathrm{Irr}\,(G) \\ \chi\,\mathrm{odd}}} L_{S_\infty}(L/K,\chi,0)^{\chi(1)} \sim \frac{h_{L,p}^-}{w_{L,p}}$$

by Proposition 2.2.1. For $\mathfrak{P} \in T$ we compute

$$\begin{aligned}
\prod_{\substack{\chi \in \mathrm{Irr}\,(G) \\ \chi\,\mathrm{odd}}} \det(1 - \phi_{\mathfrak{P}}^{-1} q_{\mathfrak{p}} | V_\chi^{I_{\mathfrak{P}}}) &= \det(1 - \phi_{\mathfrak{P}}^{-1} q_{\mathfrak{p}} | \bigoplus_{\chi\,\mathrm{odd}} \chi(1) V_\chi^{I_{\mathfrak{P}}}) \\
&= \det(1 - \phi_{\mathfrak{P}}^{-1} q_{\mathfrak{p}} | \mathbb{C}[G/I_{\mathfrak{P}}]^-) \\
&= \det(1 - \phi_{\mathfrak{P}}^{-1} q_{\mathfrak{p}} | \mathbb{Z}[G/I_{\mathfrak{P}}]^-) \\
&\sim |\mathbb{Z}_p[G/I_{\mathfrak{P}}]^- / 1 - \phi_{\mathfrak{P}}^{-1} q_{\mathfrak{p}}| \\
&= |\mathbb{Z}_p[G/I_{\mathfrak{P}}]^- / q_{\mathfrak{p}} - \phi_{\mathfrak{P}}| \\
&\overset{(1)}{=} |(\mathfrak{o}_L / \prod_{g \in G/G_{\mathfrak{P}}} \mathfrak{P}^g)^{\times,-}|_p.
\end{aligned}$$

Here, equation (1) derives from the exact sequence

$$\mathbb{Z}_p[G/I_{\mathfrak{P}}] \rightarrowtail \mathbb{Z}_p[G/I_{\mathfrak{P}}] \twoheadrightarrow (\mathfrak{o}_L / \prod_{g \in G/G_{\mathfrak{P}}} \mathfrak{P}^g)^\times \otimes \mathbb{Z}_p,$$

where the first map is $1 \mapsto q_{\mathfrak{p}} - \phi_{\mathfrak{P}}$ and the second sends 1 to a generator of $(\mathfrak{o}_L/\mathfrak{P})^\times$. Since $j \in G_{\mathfrak{P}}$ for all primes $\mathfrak{P} \in S_1$, we have

$$\prod_{\substack{\chi \in \mathrm{Irr}\,(G) \\ \chi\,\mathrm{odd}}} \det(1 - \phi_{\mathfrak{P}}^{-1} | V_\chi^{I_{\mathfrak{P}}} / V_\chi^{G_{\mathfrak{P}}}) \sim 1.$$

Indeed, if actually $j \in I_{\mathfrak{P}}$, the determinant equals 1. Otherwise it is a product of some $1 - \zeta_{2m}$, where ζ_{2m} are roots of unity of even order, and hence relatively prime to p. Thus, we get

$$\prod_{\substack{\chi \in \mathrm{Irr}\,(G) \\ \chi\,\mathrm{odd}}} (\Theta_{S_1}^T(\chi))^{\chi(1)} \sim \frac{h_{L,p}^-}{w_{L,p}} \prod_{\mathfrak{P} \in T^*} |(\mathfrak{o}_L / \prod_{g \in G/G_{\mathfrak{P}}} \mathfrak{P}^g)^{\times,-}|_p = |A_L^T|_p$$

by (2.9).

Now let G be abelian. If $\frac{1-j}{2}\theta_{S_1}^T \in \mathbb{Z}_pG^-$, the left hand side of the above equation equals $|(\mathbb{Z}_pG)_-/\theta_{S_1}^T(\mathbb{Z}_pG)_-|$. Finally, the integrality of $\frac{1-j}{2}\theta_{S_1}^T$ follows from [Ca] p. 49. More precisely, define for each prime $\mathfrak{P}$ a local module $M_\mathfrak{P}$ by

$$M_\mathfrak{P} = \langle N_{I_\mathfrak{P}}, 1 - |I_\mathfrak{P}|^{-1}N_{I_\mathfrak{P}}\phi_\mathfrak{P}^{-1}\rangle_{\mathbb{Z}I_\mathfrak{P}} \subset \mathbb{Q}I_\mathfrak{P}. \qquad (2.10)$$

Let $\mathfrak{A} = \mathrm{Ann}_{\mathbb{Z}G}(\mu_L)$ be the annihilator of the roots of unity in L. In [Gr3] the author defines the Sinnott-Kurihara ideal to be

$$SKu(L/K) = \mathfrak{A} \prod_{\mathfrak{P}\in S_{\mathrm{ram}}^*} M_\mathfrak{P} \cdot \omega\mathbb{Z}G \subset \mathbb{Z}G.$$

The proof of Proposition 2.2.5 gets completed by means of the following

Lemma 2.2.6 *Fix a setting (∗) and let G be abelian. Then*

$$\frac{1-j}{2}\theta_{S_1}^T \in SKu(L/K)^- \cdot \mathbb{Z}_pG.$$

PROOF. We have

$$\frac{1-j}{2}\theta_{S_1}^T = \frac{1-j}{2}\omega \prod_{\mathfrak{P}\in T^*}(1 - |I_\mathfrak{P}|^{-1}N_{I_\mathfrak{P}}\phi_\mathfrak{P}^{-1}q_\mathfrak{p}) \prod_{\mathfrak{P}\in S_1^*}(1 - |I_\mathfrak{P}|^{-1}N_{I_\mathfrak{P}}\phi_\mathfrak{P}^{-1}).$$

The condition on the prime $\mathfrak{P}_0 \in T$ causes $1 - N_{I_{\mathfrak{P}_0}}\phi_{\mathfrak{P}_0}^{-1}q_{\mathfrak{p}_0} \in \mathfrak{A}$. Let $\mathfrak{P} \in S_{\mathrm{ram}}^* \cap T^*$ and $q \in \mathbb{Z}$ the rational prime below $\mathfrak{P}$. If we denote the q-Sylow subgroup of the inertia group $I_\mathfrak{P}$ by $R_\mathfrak{P}$, the intermediate extension corresponding to $G_\mathfrak{P}/R_\mathfrak{P}$ is tame at $\mathfrak{P}$. Therefore, by Lemma 2.0.14, the ramification index $e_\mathfrak{P} = |I_\mathfrak{P}|$ divides $q_\mathfrak{p} - 1$ up to a power of q, since G is abelian. Hence

$$1 - |I_\mathfrak{P}|^{-1}N_{I_\mathfrak{P}}\phi_\mathfrak{P}^{-1}q_\mathfrak{p} = 1 - |I_\mathfrak{P}|^{-1}N_{I_\mathfrak{P}}\phi_\mathfrak{P}^{-1} - \phi_\mathfrak{P}^{-1}\frac{q_\mathfrak{p}-1}{e_\mathfrak{P}}N_{I_\mathfrak{P}} \in M_\mathfrak{P} \cdot \mathbb{Z}_pG.$$

For the tamely ramified primes above p the element

$$e_\mathfrak{P} = (e_\mathfrak{P} - N_{I_\mathfrak{P}})(1 - |I_\mathfrak{P}|^{-1}N_{I_\mathfrak{P}}\phi_\mathfrak{P}^{-1}) + N_{I_\mathfrak{P}} \in M_\mathfrak{P}$$

lies in $\mathbb{Z}_pG^\times$, since $p \nmid e_\mathfrak{P}$. Therefore, we get $M_\mathfrak{P} \cdot \mathbb{Z}_pG = \mathbb{Z}_pG$ in this case. Finally, we obviously have $(1 - |I_\mathfrak{P}|^{-1}N_{I_\mathfrak{P}}\phi_\mathfrak{P}^{-1}) \in M_\mathfrak{P}$ for the primes $\mathfrak{P} \in S_1$. □

In the next section we are going to show that the minus part of the LRNC for L/K at $p \neq 2$ can be restated in terms of a representing homomorphism for $A_L^T \otimes \mathbb{Z}_p$. The homomorphism involved is just the image of $\Theta_{S_1}^T$ in $\mathrm{Hom}_{\Gamma_\wp}(R_p^-(G), F_\wp^\times)$. Hence, Proposition 2.2.5 will give some evidence of the conjecture by means of the following

Proposition 2.2.7 *Let G be a finite group, p a finite rational prime and $R_p = \mathbb{Z}_pG$ (or $R_p = \mathbb{Z}_pG_+$, $\mathbb{Z}_pG_-$ if $p \neq 2$). If a finite c.t. R_p-module A has representing homomorphism $\chi \mapsto f(\chi)$, there exists an $\alpha \in \mathbb{Z}_p^\times$ such that*

$$|A| = \alpha \cdot \prod_{\chi \in \mathrm{Irr}\,(G)} f(\chi)^{\chi(1)},$$

where we set $f(\chi) = 1$ if $R_p = \mathbb{Z}_pG_+$ and χ is odd or if $R_p = \mathbb{Z}_pG_-$ and χ is even.

PROOF. We only treat the case where $R_p = \mathbb{Z}_pG$; the others are similar. Since $|\cdot|$ is multiplicative on short exact sequences of finite modules, we get a well defined map

$$|\cdot| : K_0T(\mathbb{Z}_pG) \to \mathbb{Z}.$$

Since a c.t. $\mathbb{Z}_pG$-module has projective dimension at most 1, there is an injection $\phi : \mathbb{Z}_pG^n \rightarrowtail \mathbb{Z}_pG^n$ such that $A = \operatorname{cok}\phi$.
Choose a local number field $F_\wp$, Galois over $\mathbb{Q}_p$ with Galois group $\Gamma_\wp$, which is large enough such that all representations of G can be realized over $F_\wp$. Then $\operatorname{cok}\phi$ has representing homomorphism

$$\chi \mapsto \det(\phi|\mathrm{Hom}_{\Gamma_\wp}(V_\chi, F_\wp G^n)).$$

We compute

$$\begin{aligned}
\prod_{\chi \in \mathrm{Irr}\,(G)} \det(\phi|\mathrm{Hom}_{\Gamma_\wp}(V_\chi, F_\wp G^n))^{\chi(1)} &= \det(\phi|\mathrm{Hom}_{\Gamma_\wp}(\bigoplus_{\chi \in \mathrm{Irr}\,(G)} \chi(1)V_\chi, F_\wp G^n)) \\
&= \det(\phi|\mathrm{Hom}_{\Gamma_\wp}(F_\wp G, F_\wp G^n)) \\
&= \det(\phi|F_\wp G^n) \\
&= \det(\phi|\mathbb{Z}_pG^n) \\
&= \alpha \cdot |\operatorname{cok}\phi|
\end{aligned}$$

with an appropriate element $\alpha \in \mathbb{Z}_p^\times$. □

REMARK. If G is abelian, the elements in $K_0T(R_p)$ can be described in terms of Fitting ideals. In this context Proposition 2.2.7 simply repeats the well known fact that

$$|A| = |R_p/\mathrm{Fitt}_{R_p}(A)|$$

for each finite c.t. R_p-module A.

2.3 A restatement of the LRNC on minus parts

The aim of this section is to prove

Theorem 2.3.1 *Fix a setting (*), where*

$$T = S_{\mathrm{ram}} \setminus (S_{\mathrm{ram}} \cap S_p) \cup \{\mathfrak{P}_0^g : g \in G\}.$$

Then $\Theta_{S_1}^T \in \mathrm{Hom}_{\Gamma_\wp}(R_p^-(G), F_\wp^\times)$ *is the representing homomorphism of the class of* $A_L^T \otimes \mathbb{Z}_p$ *in* $K_0T(\mathbb{Z}_pG_-)$ *if and only if the minus part of the LRNC at p holds for* L/K.

Once again, it seems to be unavoidable to go through the construction of Tate-sequences. This time we choose a set S of places of L which is small in the sense that S contains no ramified primes. More precisely, we choose $S = S_f \cup S_\infty$, where S_f is a set of totally decomposed primes such that the ray class group cl_L^T is generated by these primes and $S_f \cap T = \emptyset$. Hence, $\mathbb{Z}S_f$ is $\mathbb{Z}G$-free of rank $s^* = |S_f^*|$ and sequence (2.2) reads

$$E_{S_\infty}^T \rightarrowtail E_S^T \to \mathbb{Z}S_f \twoheadrightarrow \mathrm{cl}_L^T.$$

In particular, the S-class group cl_S is trivial, and $\nabla_S = \overline{\nabla}_S$.

Tensoring with $\mathbb{Z}_p$ and taking minus parts of the above sequence gives

$$E_S^{T,-} \otimes \mathbb{Z}_p \rightarrowtail \mathbb{Z}_pS^- \twoheadrightarrow A_L^T \otimes \mathbb{Z}_p. \tag{2.11}$$

Since $\mathbb{Z}_pS^- = \mathbb{Z}_pS_f^-$ is $\mathbb{Z}_pG_-$-free and A_L^T is c.t. by Theorem 2.1.1, we have proven

Lemma 2.3.2 *The* $\mathbb{Z}_pG_-$*-module* $E_S^{T,-} \otimes \mathbb{Z}_p$ *is cohomologically trivial.*

Let $\mathfrak{P}$ be a finite prime of L. Take an exact sequence

$$L_\mathfrak{P}^\times \rightarrowtail V_\mathfrak{P} \twoheadrightarrow \Delta G_\mathfrak{P}$$

whose extension class in $\mathrm{Ext}_{G_\mathfrak{P}}^1(\Delta G_\mathfrak{P}, L_\mathfrak{P}^\times) \simeq H^2(G_\mathfrak{P}, L_\mathfrak{P}^\times)$ is the local fundamental class of $L_\mathfrak{P}/K_\mathfrak{p}$. Recall that the inertial lattice $W_\mathfrak{P}$ is the push-out along the normalized valuation $v_\mathfrak{P} : L_\mathfrak{P}^\times \twoheadrightarrow \mathbb{Z}$ (cf. diagram (1.24)). We are going to repeat this process once more.
We have exact sequences

$$U_\mathfrak{P} \rightarrowtail V_\mathfrak{P} \twoheadrightarrow W_\mathfrak{P},$$

$$U_\mathfrak{P}^1 \rightarrowtail U_\mathfrak{P} \twoheadrightarrow (\mathfrak{o}_L/\mathfrak{P})^\times$$

and define $T_{\mathfrak{P}}$ to be the push-out of the upper sequence along the projection of the lower sequence as shown in the following commutative diagram

$$\begin{array}{ccccc} U^1_{\mathfrak{P}} & = & U^1_{\mathfrak{P}} & & \\ \downarrow & & \downarrow & & \\ U_{\mathfrak{P}} & \hookrightarrow & V_{\mathfrak{P}} & \twoheadrightarrow & W_{\mathfrak{P}} \\ \downarrow & & \downarrow & & \| \\ (\mathfrak{o}_L/\mathfrak{P})^\times & \hookrightarrow & T_{\mathfrak{P}} & \twoheadrightarrow & W_{\mathfrak{P}} \end{array} \qquad (2.12)$$

Lemma 2.3.3 *(1) The G-module $\mathrm{ind}^G_{G_{\mathfrak{P}}} T_{\mathfrak{P}} \otimes \mathbb{Z}_p$ is cohomologically trivial for each finite prime $\mathfrak{P} \nmid p$ of L and for each finite prime $\mathfrak{P}$ which is at most tamely ramified in L/K.*

(2) The G-module $(\mathrm{ind}^G_{G_{\mathfrak{P}}} T_{\mathfrak{P}})^- \otimes \mathbb{Z}_p$ is cohomologically trivial for each finite prime $\mathfrak{P} \mid p$.

PROOF. Let P be a p-Sylow subgroup of G. We denote the p-completion of any module M by $\widehat{M}$; especially, if M is finitely generated as $\mathbb{Z}$-module, we have $\widehat{M} = M \otimes \mathbb{Z}_p$.
We start with the case $\mathfrak{P} \nmid p$. Then $\widehat{U^1_{\mathfrak{P}}}$ vanishes, since $U^1_{\mathfrak{P}}$ is a pro-q-group for a prime $q \neq p$. The exact sequence

$$U^1_{\mathfrak{P}} \rightarrowtail V_{\mathfrak{P}} \twoheadrightarrow T_{\mathfrak{P}}$$

now implies that for all $i \in \mathbb{Z}$ we have

$$H^i(P, \mathrm{ind}^G_{G_{\mathfrak{P}}} T_{\mathfrak{P}} \otimes \mathbb{Z}_p) = H^i(P_{\mathfrak{P}}, T_{\mathfrak{P}} \otimes \mathbb{Z}_p) \simeq H^i(P_{\mathfrak{P}}, \widehat{V_{\mathfrak{P}}}) = 1,$$

since $\widehat{V_{\mathfrak{P}}}$ is c.t. by [GW], p. 282.
Now let $\mathfrak{P}$ be a prime above p. Then the bottom sequence of diagram (2.12) implies that $T_{\mathfrak{P}} \otimes \mathbb{Z}_p = W_{\mathfrak{P}} \otimes \mathbb{Z}_p$. The canonical projection $G_{\mathfrak{P}} \twoheadrightarrow \overline{G_{\mathfrak{P}}}$ induces an exact sequence

$$\Delta(G_{\mathfrak{P}}, I_{\mathfrak{P}}) \rightarrowtail \mathbb{Z}G_{\mathfrak{P}} \twoheadrightarrow \mathbb{Z}\overline{G_{\mathfrak{P}}}.$$

The projection onto the second component of $W_{\mathfrak{P}} \subset \Delta G_{\mathfrak{P}} \times \mathbb{Z}\overline{G_{\mathfrak{P}}}$ yields a quite similar exact sequence

$$\Delta(G_{\mathfrak{P}}, I_{\mathfrak{P}}) \rightarrowtail W_{\mathfrak{P}} \twoheadrightarrow \mathbb{Z}\overline{G_{\mathfrak{P}}}.$$

If $\mathfrak{P}$ is at most tamely ramified in L/K, the $G_{\mathfrak{P}}$-module $\mathbb{Z}_p\overline{G_{\mathfrak{P}}}$ is projective, since the corresponding idempotent lies in $\mathbb{Z}_p G_{\mathfrak{P}}$. Therefore, the p-completed versions of the above two sequences show that $W_{\mathfrak{P}} \otimes \mathbb{Z}_p \simeq \mathbb{Z}_p G_{\mathfrak{P}}$. In particular, $W_{\mathfrak{P}} \otimes \mathbb{Z}_p$ and hence $T_{\mathfrak{P}} \otimes \mathbb{Z}_p$ are c.t. $G_{\mathfrak{P}}$-modules.

We are left with the case $\mathfrak{P} \mid p$ and $j \in G_{\mathfrak{P}}$. Then j already acts on $G_{\mathfrak{P}}$-modules, and the two exact sequences

$$\mathbb{Z} \rightarrowtail W_{\mathfrak{P}} \twoheadrightarrow \Delta G_{\mathfrak{P}}, \quad \Delta G_{\mathfrak{P}} \rightarrowtail \mathbb{Z}G_{\mathfrak{P}} \twoheadrightarrow \mathbb{Z}$$

imply that $T_{\mathfrak{P}}^- \otimes \mathbb{Z}_p = W_{\mathfrak{P}}^- \otimes \mathbb{Z}_p \simeq \mathbb{Z}_p G_{\mathfrak{P}}^-$, since $\mathbb{Z}^-$ and likewise $\mathbb{Z}_p^-$ are zero. This completes the proof. □

As required for the construction of Tate-sequences, we now choose a finite set S' of places of L which contains $S \cup S_{\mathrm{ram}}$ and is large enough to generate the ideal class group of L, and such that $\bigcup_{\mathfrak{P} \in S'} G_{\mathfrak{P}} = G$. In addition, we may assume that $T \subset S'$. We set

$$T_{S'} = \bigoplus_{\mathfrak{P} \in T^*} \operatorname{ind}_{G_{\mathfrak{P}}}^G T_{\mathfrak{P}} \oplus \bigoplus_{\mathfrak{P} \in S^*} \operatorname{ind}_{G_{\mathfrak{P}}}^G \Delta G_{\mathfrak{P}} \oplus \bigoplus_{\mathfrak{P} \in S'^* \setminus (S^* \cup T^*)} \operatorname{ind}_{G_{\mathfrak{P}}}^G W_{\mathfrak{P}}.$$

Let $\mathfrak{M}_T = \prod_{\mathfrak{P} \in T} \mathfrak{P}$ as before, and define an idèle subgroup

$$J_S^T := \prod_{\mathfrak{P} \in T} U_{\mathfrak{P}}^1 \times \prod_{\mathfrak{P} \in S} L_{\mathfrak{P}}^\times \times \prod_{\mathfrak{P} \notin S \cup T} U_{\mathfrak{P}}^0.$$

The diagrams (2.12) for $\mathfrak{P} \in T$ together with the first step in the construction of Tate-sequences give rise to the commutative diagram

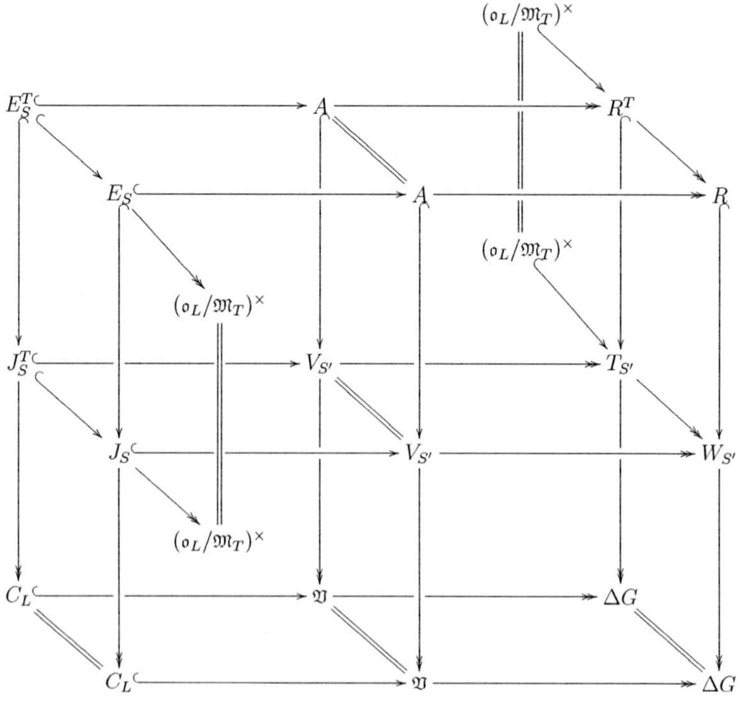

If we take the direct sum of the exact sequences

$$\begin{array}{rll} \Delta G_{\mathfrak{P}} \rightarrowtail \mathbb{Z}G_{\mathfrak{P}} \twoheadrightarrow \mathbb{Z} & \text{for} & \mathfrak{P} \in S^* \\ W_{\mathfrak{P}} \rightarrowtail \mathbb{Z}G_{\mathfrak{P}}^2 \twoheadrightarrow W_{\mathfrak{P}}^0 & \text{for} & \mathfrak{P} \in (S_{\text{ram}} \cap S_p)^* \\ T_{\mathfrak{P}} \rightarrowtail T_{\mathfrak{P}} \oplus \mathbb{Z}G_{\mathfrak{P}}^2 \twoheadrightarrow \mathbb{Z}G_{\mathfrak{P}}^2 & \text{for} & \mathfrak{P} \in (T \cap S_{\text{ram}})^* \\ T_{\mathfrak{P}} \overset{=}{\rightarrowtail} T_{\mathfrak{P}} \twoheadrightarrow 0 & \text{for} & \mathfrak{P} = \mathfrak{P}_0 \\ W_{\mathfrak{P}} \overset{\simeq}{\rightarrowtail} \mathbb{Z}G_{\mathfrak{P}} \twoheadrightarrow 0 & \text{for} & \mathfrak{P} \in (S' \setminus (S \cup S_{\text{ram}} \cup T))^*, \end{array}$$

we get an exact sequence

$$T_{S'} \rightarrowtail N_{S'}^T \twoheadrightarrow M_*^T,$$

where $N_{S'}^T$ and M_*^T are the direct sums of the middle and the right-hand modules of the above sequences.

Note that the exact sequence

$$W_{S'} \rightarrowtail N_{S'} \twoheadrightarrow M^*$$

of diagram (1.31) derives from a similar construction. We have only modified the exact sequences for the primes $\mathfrak{P} \in T^*$. The relation is comprised in the following two obviously commutative diagrams:

$$\begin{array}{ccc} (\mathfrak{o}_L/\mathfrak{P}_0)^\times & = & (\mathfrak{o}_L/\mathfrak{P}_0)^\times \\ \downarrow & & \downarrow \\ T_{\mathfrak{P}_0} & = & T_{\mathfrak{P}_0} \\ \downarrow & & \downarrow \\ W_{\mathfrak{P}_0} & \xrightarrow{\simeq} & \mathbb{Z}G_{\mathfrak{P}_0} \end{array}$$

for the prime $\mathfrak{P}_0 \in T^*$, and

$$\begin{array}{ccccc} (\mathfrak{o}_L/\mathfrak{P})^\times & \xrightarrow{-\text{id}} & T_{\mathfrak{P}} & \twoheadrightarrow & W_{\mathfrak{P}} \\ \downarrow & & \downarrow{\scriptstyle -\text{id}\oplus\tau_{\mathfrak{P}}} & & \downarrow \\ T_{\mathfrak{P}} & \hookrightarrow & T_{\mathfrak{P}} \oplus \mathbb{Z}G_{\mathfrak{P}}^2 & \twoheadrightarrow & \mathbb{Z}G_{\mathfrak{P}}^2 \\ \downarrow{\scriptstyle t_{\mathfrak{P}}} & & \downarrow{\scriptstyle \tau_{\mathfrak{P}}\oplus\text{id}} & & \downarrow \\ W_{\mathfrak{P}} & \hookrightarrow & \mathbb{Z}G_{\mathfrak{P}}^2 & \twoheadrightarrow & W_{\mathfrak{P}}^0 \end{array}$$

for the primes $\mathfrak{P} \in (T \cap S_{\text{ram}})^*$, where the map $\tau_{\mathfrak{P}} : T_{\mathfrak{P}} \to \mathbb{Z}G_{\mathfrak{P}}^2$ is the composition of the surjection $t_{\mathfrak{P}} : T_{\mathfrak{P}} \twoheadrightarrow W_{\mathfrak{P}}$ and the inclusion $W_{\mathfrak{P}} \rightarrowtail \mathbb{Z}G_{\mathfrak{P}}^2$.

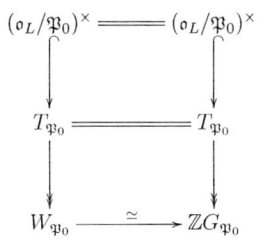

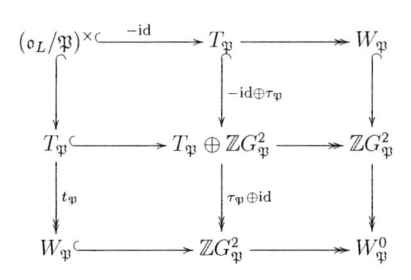

Hence, we get the commutative diagram

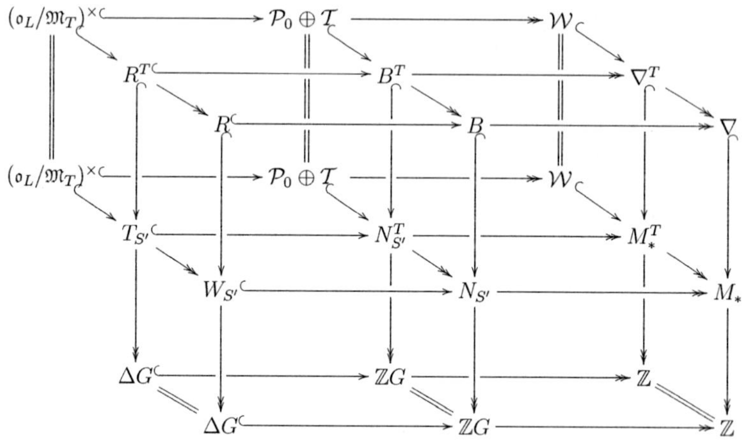

where we have defined

$$\begin{aligned} \mathcal{P}_0 &= \operatorname{ind}_{G_{\mathfrak{P}_0}}^G (\mathfrak{o}_L/\mathfrak{P}_0)^\times, \\ \mathcal{T} &= \bigoplus_{\mathfrak{P}\in(S_{\mathrm{ram}}\cap T)^*} \operatorname{ind}_{G_{\mathfrak{P}}}^G T_{\mathfrak{P}}, \\ \mathcal{W} &= \bigoplus_{\mathfrak{P}\in(S_{\mathrm{ram}}\cap T)^*} \operatorname{ind}_{G_{\mathfrak{P}}}^G W_{\mathfrak{P}}. \end{aligned}$$

The roofs of the last two three-dimensional diagrams fit together as shown in the following diagram:

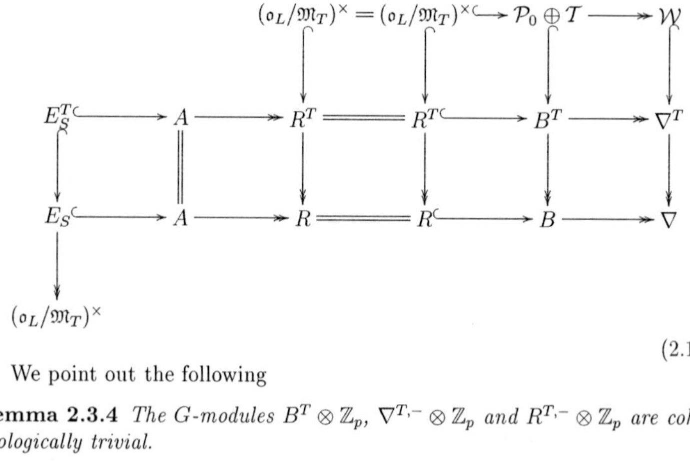

(2.13)

We point out the following

Lemma 2.3.4 *The G-modules $B^T \otimes \mathbb{Z}_p$, $\nabla^{T,-} \otimes \mathbb{Z}_p$ and $R^{T,-} \otimes \mathbb{Z}_p$ are cohomologically trivial.*

PROOF. The G-module $N_{S'}^T \otimes \mathbb{Z}_p$ is c.t. by Lemma 2.3.3 and its definition. Therefore, $B^T \otimes \mathbb{Z}_p$ is also c.t., since B^T is the kernel of $N_{S'}^T \twoheadrightarrow \mathbb{Z}G$.
Once more by Lemma 2.3.3 and the choice of the set S the module $\nabla^{T,-} \otimes \mathbb{Z}_p = M_*^{T,-} \otimes \mathbb{Z}_p$ is c.t. For this, note that $T_{\mathfrak{P}} \otimes \mathbb{Z}_p = W_{\mathfrak{P}} \otimes \mathbb{Z}_p$ for all primes $\mathfrak{P}$ above p, and that the cohomology of $W_{\mathfrak{P}}$ and $W_{\mathfrak{P}}^0$ are closely related by means of the exact sequence

$$W_{\mathfrak{P}} \rightarrowtail \mathbb{Z}G_{\mathfrak{P}}^2 \twoheadrightarrow W_{\mathfrak{P}}^0.$$

Finally, the exact sequence

$$R^T \rightarrowtail B^T \twoheadrightarrow \nabla^T$$

implies the corresponding result for $R^{T,-} \otimes \mathbb{Z}_p$. □

We now intend to define an isomorphism ϕ as required for the construction of the element Ω_ϕ. Since the cokernel of the injection $E_S^T \rightarrowtail E_S$ is finite, we can choose an injection $\phi_S^T : \Delta S \rightarrowtail E_S^T$. Hence, we get an injection ϕ_S as shown in the diagram:

$$\begin{array}{ccc} \Delta S & \xrightarrow{\phi_S^T} & E_S^T \\ & \searrow^{\phi_S} & \downarrow \\ & & E_S \end{array}$$

Recall that for each finite prime $\mathfrak{P}$ of L the element $d_{\mathfrak{P}} = |G_{\mathfrak{P}}|^{-1}\kappa(|G_{\mathfrak{P}}|, N_{G_{\mathfrak{P}}})$ is a $\mathbb{Q}G_{\mathfrak{P}}$-generator of $\mathbb{Q}W_{\mathfrak{P}}^0$. Hence, we can define isomorphisms

$$\begin{array}{rcl} \delta_{\mathfrak{P}} : \mathbb{Q}W_{\mathfrak{P}}^0 & \longrightarrow & \mathbb{Q}G_{\mathfrak{P}} \\ d_{\mathfrak{P}} & \mapsto & 1, \end{array}$$

and set $d := \sum_{\mathfrak{P} \in S_{\mathrm{ram}}^*} \operatorname{ind} \delta_{\mathfrak{P}}$. Let C be a $\mathbb{Z}G$-free module of rank $|S_{\mathrm{ram}}^*|$ with basis $1_{\mathfrak{P}}$, $\mathfrak{P} \in S_{\mathrm{ram}}^*$, and define ϕ to be the $\mathbb{Q}G$-isomorphism

$$\phi : \mathbb{Q}\nabla \simeq \mathbb{Q}(\Delta S \oplus \bigoplus\nolimits_{\mathfrak{P} \in S_{\mathrm{ram}}^*} \operatorname{ind}_{G_{\mathfrak{P}}}^G W_{\mathfrak{P}}^0) \xrightarrow{\mathbb{Q} \otimes \phi_S \oplus d} \mathbb{Q}(E_S \oplus C)$$

Here, the first isomorphism is induced by the natural inclusion on minus parts, whereas we have to choose a splitting of sequence (1.13) on plus parts (after tensoring with $\mathbb{Q}$). But this choice will play no decisive role, since we are going to deal with minus parts only.

In analogy to the elements $d_{\mathfrak{P}}$, we define $\mathbb{Q}G_{\mathfrak{P}}$-generators $c_{\mathfrak{P}}$ of $\mathbb{Q}W_{\mathfrak{P}}$ by

$$c_{\mathfrak{P}} := (1 - \frac{1}{|G_{\mathfrak{P}}|} N_{G_{\mathfrak{P}}}, N_{\overline{G_{\mathfrak{P}}}} + (\phi_{\mathfrak{P}} - 1)^{-1}(1 - \frac{1}{|\overline{G_{\mathfrak{P}}}|} N_{\overline{G_{\mathfrak{P}}}})), \qquad (2.14)$$

where $\overline{G_{\mathfrak{P}}} = G_{\mathfrak{P}}/I_{\mathfrak{P}}$ as before, and

$$(\phi_{\mathfrak{P}} - 1)^{-1} = \frac{1}{|\overline{G_{\mathfrak{P}}}|} \sum_{i=0}^{|\overline{G_{\mathfrak{P}}}|-1} i\phi_{\mathfrak{P}}^i.$$

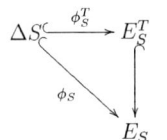

We establish a connection between the generators $c_{\mathfrak{P}}$ and $d_{\mathfrak{P}}$ by means of the commutative diagram

$$\begin{array}{ccc} \mathbb{Q}G_{\mathfrak{P}} & \xrightarrow{1 \mapsto c_{\mathfrak{P}}} & \mathbb{Q}W_{\mathfrak{P}} \\ \big\downarrow \iota_1 & & \big\downarrow q \\ \mathbb{Q}G_{\mathfrak{P}}^2 & \xrightarrow{g_{\mathfrak{P}}} & \mathbb{Q}G_{\mathfrak{P}}^2 \\ \big\downarrow \pi_2 & & \big\downarrow \kappa \\ \mathbb{Q}G_{\mathfrak{P}} & \xrightarrow{1 \mapsto d_{\mathfrak{P}}} & \mathbb{Q}W_{\mathfrak{P}}^0 \end{array}$$

where the maps of the left column are the natural inclusion into the first and the projection onto the second component. The isomorphism $g_{\mathfrak{P}}$ is defined to be

$$\begin{array}{rcl} g_{\mathfrak{P}} : \mathbb{Q}G_{\mathfrak{P}}^2 & \longrightarrow & \mathbb{Q}G_{\mathfrak{P}}^2 \\ (1,0) & \mapsto & q(c_{\mathfrak{P}}) \\ & & = (N_{G_{\mathfrak{P}}} + (\phi_{\mathfrak{P}} - 1)^{-1}(N_{I_{\mathfrak{P}}} - \frac{1}{|G_{\mathfrak{P}}|}N_{G_{\mathfrak{P}}}), \phi_{\mathfrak{P}}^{-1}(1 - \frac{1}{|G_{\mathfrak{P}}|}N_{G_{\mathfrak{P}}})) \\ (0,1) & \mapsto & (1, \frac{1}{|G_{\mathfrak{P}}|}N_{G_{\mathfrak{P}}}) \end{array}$$

Let us split the free $\mathbb{Z}G$-module C into

$$C = C_{p'} \oplus C_p,$$

where C_p is free of rank $|(S_{\mathrm{ram}} \cap S_p)^*|$. If we combine the above diagram for all primes $\mathfrak{P} \in S_{\mathrm{ram}}^*$ which do not lie above p, we get the following commutative diagram on minus parts:

$$\begin{array}{ccc} \mathcal{W}^- & \overset{c}{\dashrightarrow} & C_{p'}^- \\ \big\downarrow & & \big\downarrow \\ \nabla^{T,-} & \dashrightarrow & (E_S \oplus C_{p'}^2 \oplus C_p)^- \\ \big\downarrow & & \big\downarrow \\ \nabla^- & \overset{\phi}{\dashrightarrow} & (E_S \oplus C)^- \end{array}$$

Here, the dotted maps only exist after tensoring with $\mathbb{Q}$, and we have defined

$$c := \sum_{\mathfrak{P} \in (S_{\mathrm{ram}} \cap T)^*} \mathrm{ind}\,(c_{\mathfrak{P}} \mapsto 1_{\mathfrak{P}}).$$

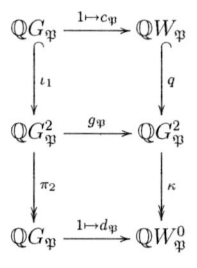

The map $g := \sum_{\mathfrak{P}\in(S_{\mathrm{ram}}\cap T)^*} \operatorname{ind} g_{\mathfrak{P}}$ is incorporated in the middle dotted arrow.

We now go into the construction of the element $\Omega_\phi^{(p),-}$ involved in the LRNC. First of all, we choose an automorphism β of $\mathbb{Q}R$ and an isomorphism $\tilde\beta$ as shown in the diagram

$$\begin{array}{ccccc}
\mathbb{Q}R & \xrightarrow{\iota} & \mathbb{Q}B & \xrightarrow{\pi} & \mathbb{Q}\nabla \\
\downarrow{\scriptstyle\beta} & & \downarrow{\scriptstyle\tilde\beta} & & \| \\
\mathbb{Q}R & \longrightarrow & \mathbb{Q}(R\oplus\nabla) & \longrightarrow & \nabla
\end{array}$$

If $\sigma : \mathbb{Q}B \to \mathbb{Q}R$ is a section of ι, we may take $\tilde\beta = \beta\sigma + \pi$. Let us tensor the righthand part of diagram (2.13) with $\mathbb{Q}$, namely

$$\begin{array}{ccccc}
\mathbb{Q}R^T & \xrightarrow{\iota^T} & \mathbb{Q}B^T & \xrightarrow{\pi^T} & \mathbb{Q}\nabla^T \\
\simeq\downarrow{\scriptstyle\pi_R} & & \downarrow{\scriptstyle\pi_B} & & \downarrow \\
\mathbb{Q}R & \xrightarrow{\iota} & \mathbb{Q}B & \xrightarrow{\pi} & \mathbb{Q}\nabla
\end{array}$$

We define a section of ι^T to be

$$\sigma^T := \pi_R^{-1}\sigma\pi_B : \mathbb{Q}B^T \to \mathbb{Q}R^T,$$

and set $\beta^T := \pi_R^{-1}\beta\pi_R$ and $\tilde\beta^T := \beta^T\sigma^T + \pi^T$ such that

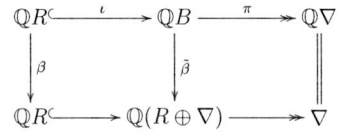

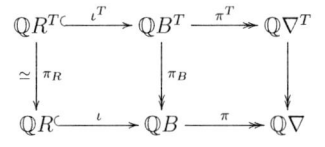

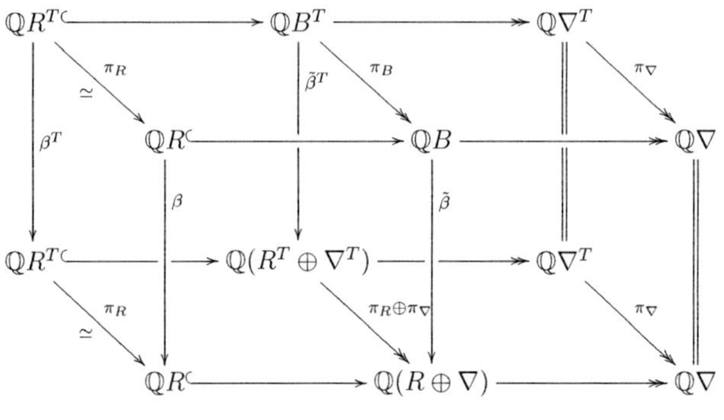

commutes. Note that

$$[\mathbb{Q}R,\beta] = [\mathbb{Q}R^T,\beta^T] \in K_1(\mathbb{Q}G). \tag{2.15}$$

Correspondingly, we choose an automorphism α of $\mathbb{Q}R$ and get a commutative diagram

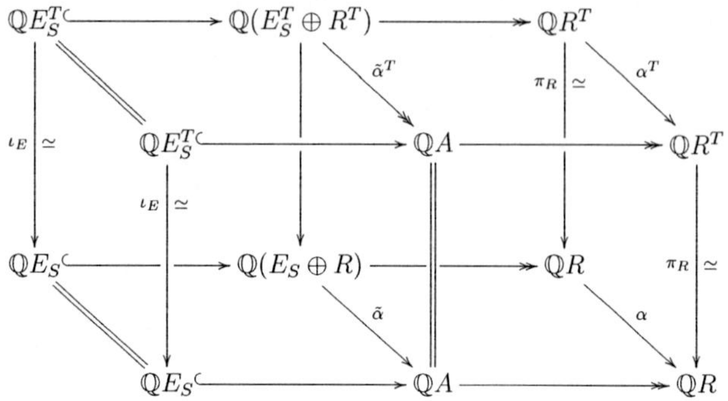

and an equality

$$[\mathbb{Q}R, \alpha] = [\mathbb{Q}R^T, \alpha^T] \in K_1(\mathbb{Q}G). \tag{2.16}$$

It turns out to be helpful to write the isomorphism $\tilde{\phi}$ defined in (1.19) in the following more complicated way.

$$\tilde{\phi}: \mathbb{Q}B^- \xrightarrow{\quad\tilde{\beta}\quad} \mathbb{Q}(R \oplus \nabla)^- \xrightarrow{\pi_R^{-1} \oplus \mathrm{id}} \mathbb{Q}(R^T \oplus \nabla)^-$$

$$\xrightarrow{\quad\simeq\quad} \mathbb{Q}(R^T \oplus \Delta S \oplus \textstyle\bigoplus_{\mathfrak{P} \in S^*_{\mathrm{ram}}} \mathrm{ind}_{G_{\mathfrak{P}}}^G W_{\mathfrak{P}}^0)^-$$

$$\xrightarrow{\quad\mathrm{id} \oplus d\quad} \mathbb{Q}(R^T \oplus \Delta S \oplus C)^- \xrightarrow{\mathrm{id} \oplus \phi_S^T \oplus \mathrm{id}} \mathbb{Q}(R^T \oplus E_S^T \oplus C)^-$$

$$\xrightarrow{\pi_R \oplus \iota_E \oplus \mathrm{id}} \mathbb{Q}(R \oplus E_S \oplus C)^- \xrightarrow{\tilde{\alpha} \oplus \mathrm{id}} \mathbb{Q}(A \oplus C)^-$$

Since $(R^T \oplus \Delta S \oplus C)^- \otimes \mathbb{Z}_p$ and $(R^T \oplus E_S^T \oplus C)^- \otimes \mathbb{Z}_p$ are c.t. G-modules by Lemma 2.3.2, Lemma 2.3.4 and the choice of the set S, we have

$$\begin{aligned}
\Omega_\phi^{(p),-} &= (B^- \otimes \mathbb{Z}_p, \tilde{\phi}, (A \oplus C)^- \otimes \mathbb{Z}_p) - \partial[\mathbb{Q}_p R^-, \alpha\beta] \\
&= (B^- \otimes \mathbb{Z}_p, (\mathrm{id} \oplus d)(\pi_R^{-1} \oplus \mathrm{id})\tilde{\beta}, (R^T \oplus \Delta S \oplus C)^- \otimes \mathbb{Z}_p) \\
&\quad + i_G(\operatorname{cok} \phi_S^T \otimes \mathbb{Z}_p) \\
&\quad + ((R^T \oplus E_S^T)^- \otimes \mathbb{Z}_p, \tilde{\alpha}(\pi_R \oplus \iota_E), A^- \otimes \mathbb{Z}_p) \\
&\quad - \partial[\mathbb{Q}_p R^-, \alpha\beta]
\end{aligned} \tag{2.17}$$

Note that the G-module $\operatorname{cok} \phi_S^T \otimes \mathbb{Z}_p$ is c.t. and finite, and therefore defines an element in $K_0T(\mathbb{Z}_pG)$ which is isomorphic to $K_0(\mathbb{Z}_pG, \mathbb{Q}_p)$ via the p-adic version of the isomorphism i_G defined in (1.5), which we also denote by i_G.

Since $\tilde{\alpha}(\pi_R \oplus \iota_E) = \tilde{\alpha}^T$, equation (2.16) and the corresponding diagram prior to it imply

$$\begin{aligned} \partial[\mathbb{Q}_p R^-, \alpha] &= \partial[\mathbb{Q}_p R^{T,-}, \alpha^T] \\ &= ((R^T \oplus E_S^T)^- \otimes \mathbb{Z}_p, \tilde{\alpha}^T, A^- \otimes \mathbb{Z}_p) \\ &= ((R^T \oplus E_S^T)^- \otimes \mathbb{Z}_p, \tilde{\alpha}(\pi_R \oplus \iota_E), A^- \otimes \mathbb{Z}_p). \end{aligned}$$

Thus, equation (2.17) reduces to

$$\begin{aligned} \Omega_\phi^{(p),-} = {} & (B^- \otimes \mathbb{Z}_p, (\mathrm{id} \oplus d)(\pi_R^{-1} \oplus \mathrm{id})\tilde{\beta}, (R^T \oplus \Delta S \oplus C)^- \otimes \mathbb{Z}_p) \\ & + i_G(\operatorname{cok} \phi_S^T \otimes \mathbb{Z}_p) - \partial[\mathbb{Q}_p R^-, \beta]. \end{aligned} \tag{2.18}$$

For a better understanding of the first summand we make use of the following commutative diagram in which the dotted maps only exist (and are isomorphisms) after tensoring with $\mathbb{Q}_p$; we have also invisibly taken minus parts:

$$\begin{array}{ccccc}
\mathcal{P}_0 \oplus \mathcal{T} & \hookrightarrow & B^T & \twoheadrightarrow & B \\
\downarrow t & & \downarrow \tilde{\beta}^T & & \downarrow \tilde{\beta} \\
(\mathfrak{o}_L/\mathfrak{M}_T)^\times \oplus \mathcal{W} & \hookrightarrow & R^T \oplus \nabla^T & \xrightarrow{\pi_R \oplus \pi_\nabla} & R \oplus \nabla \\
\downarrow & & \| & & \downarrow \pi_R^{-1} \oplus \mathrm{id} \\
\mathcal{W} & \hookrightarrow & R^T \oplus \nabla^T & \xrightarrow{\mathrm{id} \oplus \pi_\nabla} & R^T \oplus \nabla \\
\| & & \downarrow \simeq & & \downarrow \simeq \\
\mathcal{W} & \hookrightarrow & R^T \oplus \bigoplus_{\mathfrak{P} \in (T \cap S_{\mathrm{ram}})^*} \operatorname{ind} \mathbb{Z} G_{\mathfrak{P}}^2 \oplus \bigoplus_{\mathfrak{P} \in (S_{\mathrm{ram}} \cap S_p)^*} \operatorname{ind} W_{\mathfrak{P}}^0 \oplus \Delta S & \twoheadrightarrow & R^T \oplus \bigoplus_{\mathfrak{P} \in S_{\mathrm{ram}}^*} \operatorname{ind} W_{\mathfrak{P}}^0 \oplus \Delta S \\
\downarrow c & & \downarrow \mathrm{id} \oplus g^{-1} \oplus d_p \oplus \mathrm{id} & & \downarrow \mathrm{id} \oplus d \oplus \mathrm{id} \\
C_{p'} & \hookrightarrow & R^T \oplus C_{p'}^2 \oplus C_p \oplus \Delta S & \twoheadrightarrow & R^T \oplus C \oplus \Delta S
\end{array}$$

The isomorphism $t : \mathbb{Q}_p \mathcal{T} \simeq \mathbb{Q}_p \mathcal{W}$ is induced by the projection $\mathcal{T} \twoheadrightarrow \mathcal{W}$ which appears in diagram (2.13). Note that the direct summands $\mathcal{P}_0$ and $(\mathfrak{o}_L/\mathfrak{M}_T)^\times$ vanish after tensoring with $\mathbb{Q}_p$. The map d_p is the restriction of d to $\bigoplus_{\mathfrak{P} \in (S_{\mathrm{ram}} \cap S_p)^*} \operatorname{ind} W_{\mathfrak{P}}^0$.

By Lemma 1.1.6, the above diagram implies that the first summand of the righthand side of equation (2.18) equals

$$\begin{aligned} & (B^{T,-} \otimes \mathbb{Z}_p, (\mathrm{id}_{R^{T,-}} \oplus g^{-1} \oplus d_p \oplus \mathrm{id}_{\Delta S^-})\tilde{\beta}^T, (R^T \oplus C_{p'}^2 \oplus C_p \oplus \Delta S)^- \otimes \mathbb{Z}_p) \\ & -((\mathcal{P}_0 \oplus \mathcal{T})^- \otimes \mathbb{Z}_p, ct, C_{p'}^- \otimes \mathbb{Z}_p) \end{aligned}$$

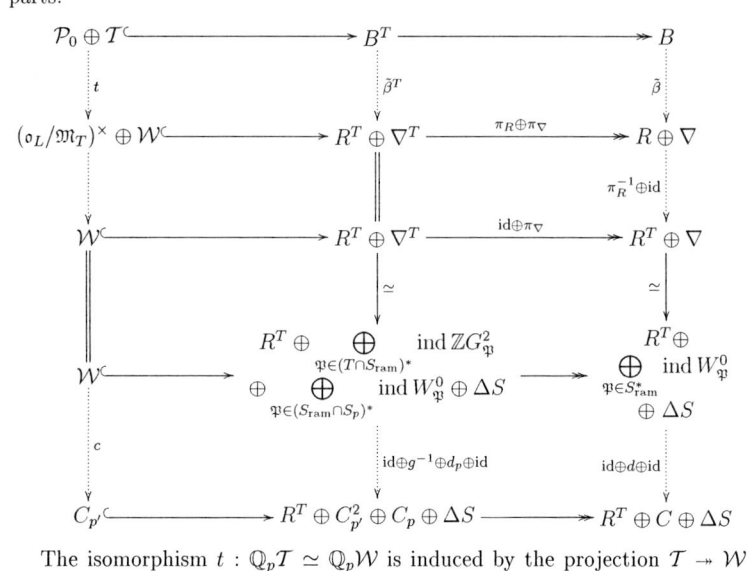

$$\begin{aligned}
&\overset{(1)}{=} \partial[\mathbb{Q}_p R^{T,-}, \beta^T] + (\nabla^{T,-} \otimes \mathbb{Z}_p, g^{-1} \oplus d_p \oplus \mathrm{id}_{\Delta S^-}, (C_{p'}^2 \oplus C_p \oplus \Delta S)^- \otimes \mathbb{Z}_p) \\
&\quad + i_G(\mathcal{P}_0^- \otimes \mathbb{Z}_p) \\
&\quad - \sum_{\mathfrak{P} \in (S_{\mathrm{ram}} \cap T)^*} ((\mathrm{ind}\, T_{\mathfrak{P}})^- \otimes \mathbb{Z}_p, \mathrm{ind}\,(c_{\mathfrak{P}} \mapsto 1_{\mathfrak{P}}) t_{\mathfrak{P}}, \mathrm{ind}\, \mathbb{Z}_p G_{\mathfrak{P}}^-) \\
&\overset{(2)}{=} \partial[\mathbb{Q}_p R^-, \beta] + i_G(\mathcal{P}_0^- \otimes \mathbb{Z}_p) + \sum_{\mathfrak{P} \in (S_{\mathrm{ram}} \cap T)^*} \partial[\mathrm{ind}\,(\mathbb{Q}_p G_{\mathfrak{P}}^2)^-, \mathrm{ind}\, g_{\mathfrak{P}}^{-1}] \\
&\quad + \sum_{\mathfrak{P} \in (S_{\mathrm{ram}} \cap S_p)^*} ((\mathrm{ind}\, W_{\mathfrak{P}}^0)^- \otimes \mathbb{Z}_p, \mathrm{ind}\, \delta_{\mathfrak{P}}, (\mathrm{ind}\, \mathbb{Z}_p G_{\mathfrak{P}})^-) \\
&\quad - \sum_{\mathfrak{P} \in (S_{\mathrm{ram}} \cap T)^*} ((\mathrm{ind}\, T_{\mathfrak{P}})^- \otimes \mathbb{Z}_p, \mathrm{ind}\,(c_{\mathfrak{P}} \mapsto 1_{\mathfrak{P}}) t_{\mathfrak{P}}, \mathrm{ind}\, \mathbb{Z}_p G_{\mathfrak{P}}^-).
\end{aligned}$$

We need to explain the equalities (1) and (2). Due to Lemma 2.3.4, the middle column of the above diagram shows that we can isolate the term $\partial[\mathbb{Q}_p R^{T,-}, \beta^T] = (B^{T,-} \otimes \mathbb{Z}_p, \tilde{\beta}^T, (R^T \oplus \nabla^T)^- \otimes \mathbb{Z}_p)$. Since

$$((\mathcal{P}_0 \oplus \mathcal{T})^- \otimes \mathbb{Z}_p, ct, C_{p'}^- \otimes \mathbb{Z}_p) = -i_G(\mathcal{P}_0^- \otimes \mathbb{Z}_p) + (\mathcal{T}^- \otimes \mathbb{Z}_p, ct, C_{p'}^- \otimes \mathbb{Z}_p)$$

by the first remark following Lemma 1.1.6, we get (1), where we have used the definition of the maps c and t. (2) follows from (2.15) and the definition of the maps g and d_p.

Now let $\mathfrak{P} \in (S_{\mathrm{ram}} \cap S_p)^*$ be wildly ramified. Since by assumption $j \in G_{\mathfrak{P}}$ for these primes, the exact sequences

$$\mathbb{Z} \rightarrowtail \mathbb{Z}G_{\mathfrak{P}} \twoheadrightarrow \mathbb{Z}G_{\mathfrak{P}}/N_{G_{\mathfrak{P}}},$$

$$\mathbb{Z}G_{\mathfrak{P}}/N_{G_{\mathfrak{P}}} \rightarrowtail W_{\mathfrak{P}}^0 \twoheadrightarrow \mathbb{Z}$$

induce an isomorphism $\mathbb{Z}_p G_{\mathfrak{P}}^- \simeq (W_{\mathfrak{P}}^0)^- \otimes \mathbb{Z}_p$, which maps $(1-j)/2$ to $d_{\mathfrak{P}}$. All this can be extracted from the diagram of Proposition 1.5.4. Hence, the isomorphism $\delta_{\mathfrak{P}}$ derives, locally at p and on minus parts, from a $\mathbb{Z}_p G_{\mathfrak{P}}$-isomorphism. Therefore

$$((\mathrm{ind}\, W_{\mathfrak{P}}^0)^- \otimes \mathbb{Z}_p, \mathrm{ind}\, \delta_{\mathfrak{P}}, (\mathrm{ind}\, \mathbb{Z}_p G_{\mathfrak{P}})^-) = 0$$

for all wildly ramified primes above p.

Altogether, we get the following description of $\Omega_\phi^{(p),-}$:

$$\begin{aligned}
\Omega_\phi^{(p),-} &= i_G(\mathrm{cok}\, \phi_S^T \otimes \mathbb{Z}_p) + i_G(\mathcal{P}_0^- \otimes \mathbb{Z}_p) \\
&\quad + \sum_{\mathfrak{P} \in (S_{\mathrm{tram}} \cap S_p)^*} ((\mathrm{ind}\, W_{\mathfrak{P}}^0)^- \otimes \mathbb{Z}_p, \mathrm{ind}\, \delta_{\mathfrak{P}}, (\mathrm{ind}\, \mathbb{Z}_p G_{\mathfrak{P}})^-) \\
&\quad - \sum_{\mathfrak{P} \in (S_{\mathrm{ram}} \cap T)^*} \partial[\mathrm{ind}\,(\mathbb{Q}_p G_{\mathfrak{P}}^2)^-, \mathrm{ind}\, g_{\mathfrak{P}}] \\
&\quad - \sum_{\mathfrak{P} \in (S_{\mathrm{ram}} \cap T)^*} ((\mathrm{ind}\, T_{\mathfrak{P}})^- \otimes \mathbb{Z}_p, \mathrm{ind}\,(c_{\mathfrak{P}} \mapsto 1_{\mathfrak{P}}) t_{\mathfrak{P}}, \mathrm{ind}\, \mathbb{Z}_p G_{\mathfrak{P}}^-),
\end{aligned} \tag{2.19}$$

where we have defined $S_{\text{tram}} \subset S_{\text{ram}}$ to be the set of all primes of L which are tamely ramified in L/K.

The representing homomorphisms of most of these terms can be computed:

Proposition 2.3.5 *Keeping the notation of the current paragraph the following holds:*

(1) $i_G(\mathcal{P}_0^- \otimes \mathbb{Z}_p)$ *has representing homomorphism*

$$\chi \mapsto \det(q_0 - \phi_{\mathfrak{P}_0} | V_\chi),$$

where $q_0 = N(\mathfrak{p}_0)$ *and* $\mathfrak{p}_0 = \mathfrak{P}_0 \cap K$.

(2) Let $\mathfrak{P} \in (S_{\text{ram}} \cap S_p)^*$ *be at most tamely ramified in* L/K.
Then $((\operatorname{ind} W_{\mathfrak{P}}^0)^- \otimes \mathbb{Z}_p, \operatorname{ind} \delta_{\mathfrak{P}}, (\operatorname{ind} \mathbb{Z}_p G_{\mathfrak{P}})^-)$ *has representing homomorphism*

$$\chi \mapsto (-e_{\mathfrak{P}})^{-\dim V_\chi^{G_{\mathfrak{P}}}} \cdot \det(1 - \phi_{\mathfrak{P}}^{-1} | V_\chi^{I_{\mathfrak{P}}} / V_\chi^{G_{\mathfrak{P}}})^{-1},$$

where $e_{\mathfrak{P}} = |I_{\mathfrak{P}}|$ *is the ramification index of the prime* $\mathfrak{P}$ *in* L/K.

(3) Let $\mathfrak{P}$ *be any finite prime of* L. *Then* $\partial[\operatorname{ind} (\mathbb{Q}_p G_{\mathfrak{P}}^2)^-, \operatorname{ind} g_{\mathfrak{P}}]$ *has representing homomorphism*

$$\chi \mapsto (-|G_{\mathfrak{P}}|)^{\dim V_\chi^{G_{\mathfrak{P}}}}.$$

(4) Let $\mathfrak{P} \in (S_{\text{ram}} \cap T)^*$. *Then* $((\operatorname{ind} T_{\mathfrak{P}})^- \otimes \mathbb{Z}_p, \operatorname{ind} (c_{\mathfrak{P}} \mapsto 1_{\mathfrak{P}}) t_{\mathfrak{P}}, \operatorname{ind} \mathbb{Z}_p G_{\mathfrak{P}}^-)$ *has representing homomorphism*

$$\chi \mapsto (f_{\mathfrak{P}}(1 - q_{\mathfrak{p}}))^{-\dim V_\chi^{G_{\mathfrak{P}}}} \cdot \det(\frac{1 - \phi_{\mathfrak{P}}}{q_{\mathfrak{p}} - \phi_{\mathfrak{P}}} | V_\chi^{I_{\mathfrak{P}}} / V_\chi^{G_{\mathfrak{P}}}),$$

where $f_{\mathfrak{P}} = |\overline{G_{\mathfrak{P}}}|$ *is the degree of the corresponding residue field extension,* $q_{\mathfrak{p}} = N(\mathfrak{p})$ *and* $\mathfrak{p} = \mathfrak{P} \cap K$.

PROOF. Recall that $\mathcal{P}_0 = \operatorname{ind}_{G_{\mathfrak{P}_0}}^G (\mathfrak{o}_L/\mathfrak{P}_0)^\times$. Since $\mathfrak{P}_0$ is unramified in L/K, the decomposition group $G_{\mathfrak{P}_0}$ is cyclic with generator $\phi_{\mathfrak{P}_0}$, which acts as q_0 on $(\mathfrak{o}_L/\mathfrak{P}_0)^\times$. So (1) is clear.
For (2) let $\mathfrak{P} \in (S_{\text{ram}} \cap S_p)^*$ be tamely ramified. Then the idempotent $\varepsilon_{\mathfrak{P}} = e_{\mathfrak{P}}^{-1} N_{I_{\mathfrak{P}}}$ lies in $\mathbb{Z}_p G_{\mathfrak{P}}$, and we claim that we have an isomorphism

$$\begin{array}{ccc} \mathbb{Z}_p G_{\mathfrak{P}} & \xrightarrow{w_{\mathfrak{P}}} & W_{\mathfrak{P}}^0 \otimes \mathbb{Z}_p \\ 1 & \mapsto & \kappa(1 - \varepsilon_{\mathfrak{P}}, 1), \end{array}$$

where we once again identify the module $W_{\mathfrak{P}}^0$ with a certain cokernel as in Proposition 1.5.4. Indeed $w_{\mathfrak{P}}(\varepsilon_{\mathfrak{P}}) = \kappa(0,1)$ and

$$w_{\mathfrak{P}}(1 - \varepsilon_{\mathfrak{P}} + e_{\mathfrak{P}}^{-1}(\phi_{\mathfrak{P}}^{-1} - 1)\varepsilon_{\mathfrak{P}}) = \kappa(1 - \varepsilon_{\mathfrak{P}}, e_{\mathfrak{P}}^{-1}(\phi_{\mathfrak{P}}^{-1} - 1)) = \kappa(1,0).$$

Therefore, $w_{\mathfrak{P}}$ is surjective and hence bijective, since both modules are torsion free of the same rank. We have

$$\begin{aligned}((W^0_{\mathfrak{P}})^- \otimes \mathbb{Z}_p, \delta_{\mathfrak{P}}, (\mathbb{Z}_pG_{\mathfrak{P}})^-) &= -((\mathbb{Z}_pG_{\mathfrak{P}})^-, \delta_{\mathfrak{P}}^{-1}, (W^0_{\mathfrak{P}})^- \otimes \mathbb{Z}_p) \\ &= -((\mathbb{Z}_pG_{\mathfrak{P}})^-, \delta_{\mathfrak{P}}^{-1} w_{\mathfrak{P}}, (\mathbb{Z}_pG_{\mathfrak{P}})^-).\end{aligned}$$

Since $w_{\mathfrak{P}}(1 - \varepsilon_{\mathfrak{P}} + e_{\mathfrak{P}}^{-1}(\phi_{\mathfrak{P}}^{-1} - 1)\varepsilon_{\mathfrak{P}} + |G_{\mathfrak{P}}|^{-1}N_{G_{\mathfrak{P}}}) = d_{\mathfrak{P}}$, the representing homomorphism in demand is

$$\chi \mapsto \det(e_{\mathfrak{P}}^{-1}(\phi_{\mathfrak{P}}^{-1} - 1)|V_\chi^{I_{\mathfrak{P}}}/V_\chi^{G_{\mathfrak{P}}})^{-1}.$$

But the desired homomorphism differs from this by

$$[\chi \mapsto \det((-e_{\mathfrak{P}})\varepsilon_{\mathfrak{P}} + 1 - \varepsilon_{\mathfrak{P}}|V_\chi)] \in \mathrm{Det}\,((\mathbb{Z}_pG^-)^\times).$$

Hence, we have proved (2).
Now let $\mathfrak{P}$ be any finite prime of L. The map $g_{\mathfrak{P}}$ defines an element in $K_1(\mathbb{Q}_pG_{\mathfrak{P}})$, which is represented by the matrix

$$\begin{pmatrix} N_{G_{\mathfrak{P}}} + (\phi_{\mathfrak{P}} - 1)^{-1}(N_{I_{\mathfrak{P}}} - f_{\mathfrak{P}}^{-1}N_{G_{\mathfrak{P}}}) & 1 \\ 1 - |G_{\mathfrak{P}}|^{-1}N_{G_{\mathfrak{P}}} & |G_{\mathfrak{P}}|^{-1}N_{G_{\mathfrak{P}}} \end{pmatrix}$$

If we subtract $|G_{\mathfrak{P}}|^{-1}N_{G_{\mathfrak{P}}}$ times the first row from the second row and exchange the two columns, we obtain a matrix

$$\begin{pmatrix} 1 & N_{G_{\mathfrak{P}}} + (\phi_{\mathfrak{P}} - 1)^{-1}(N_{I_{\mathfrak{P}}} - f_{\mathfrak{P}}^{-1}N_{G_{\mathfrak{P}}}) \\ 0 & 1 - |G_{\mathfrak{P}}|^{-1}N_{G_{\mathfrak{P}}} - N_{G_{\mathfrak{P}}} \end{pmatrix}$$

Since we have only used matrix operations which does not affect the image in $K_0(\mathbb{Z}_pG, \mathbb{Q}_p)$, we get (3).
Finally, let $\mathfrak{P} \in (S_{\mathrm{ram}} \cap T)^*$, i.e. $\mathfrak{P}$ is a ramified prime not above p. It directly follows from the definition that $T_{\mathfrak{P}}$ is the push-out of the local fundamental class along the canonical projection $L_{\mathfrak{P}}^\times \twoheadrightarrow L_{\mathfrak{P}}^\times/U_{\mathfrak{P}}^1$ as shown in the commutative diagram

$$\begin{array}{ccccc} U_{\mathfrak{P}}^1 & = & U_{\mathfrak{P}}^1 & & \\ \downarrow & & \downarrow & & \\ L_{\mathfrak{P}}^\times & \hookrightarrow & V_{\mathfrak{P}} & \twoheadrightarrow & \Delta G_{\mathfrak{P}} \\ \downarrow & & \downarrow & & \| \\ L_{\mathfrak{P}}^\times/U_{\mathfrak{P}}^1 & \xrightarrow{\iota_{\mathfrak{P}}} & T_{\mathfrak{P}} & \twoheadrightarrow & \Delta G_{\mathfrak{P}} \end{array} \qquad (2.20)$$

We see that $\widehat{L_{\mathfrak{P}}^\times} = L_{\mathfrak{P}}^\times/U_{\mathfrak{P}}^1 \otimes \mathbb{Z}_p$ and $T_{\mathfrak{P}} \otimes \mathbb{Z}_p = \widehat{V_{\mathfrak{P}}}$. Actually before taking minus parts, $-(\mathrm{ind}\, T_{\mathfrak{P}} \otimes \mathbb{Z}_p, \mathrm{ind}\,(c_{\mathfrak{P}} \mapsto 1_{\mathfrak{P}})t_{\mathfrak{P}}, \mathrm{ind}\, \mathbb{Z}_pG_{\mathfrak{P}})$ is induced by applying the Ω-construction to the two-extension

$$\widehat{L_{\mathfrak{P}}^\times} \rightarrowtail \widehat{V_{\mathfrak{P}}} \to \mathbb{Z}_pG \twoheadrightarrow \mathbb{Z}_p$$

and an isomorphism $\mathbb{Q}_p \to \mathbb{Q}_p\widehat{L_{\mathfrak{P}}^{\times}}$ which maps 1 to an element that has $\mathfrak{P}$-adic valuation equal to 1. Therefore, (4) follows from Theorem D in [RW2]. If $\mathfrak{P}$ is at most tamely ramified in L/K, we can alternatively use Theorem 4.3, p. 563 in [BB]. □

Now we have computed all the representing homomorphisms for the terms of the right hand side of equation (2.19) apart from $i_G(\operatorname{cok}\phi_S^T \otimes \mathbb{Z}_p)$.
Due to the choice of the set S, we can fix an isomorphism

$$\rho_S : \Delta S^- \xrightarrow{\simeq} (\mathbb{Z}G_-)^{s^*}.$$

We build the following commutative diagram which defines a monomorphism ψ:

$$\begin{array}{ccccc}
\Delta S^- & \xrightarrow[\simeq]{\rho_S} & (\mathbb{Z}G_-)^{s^*} & & \\
\downarrow \phi_S^T & & \downarrow \psi & & \\
E_S^{T,-} & \hookrightarrow & (\mathbb{Z}G_-)^{s^*} & \twoheadrightarrow & A_L^T \\
\downarrow & & \downarrow & & \| \\
\operatorname{cok}\phi_S^T & \hookrightarrow & \operatorname{cok}\psi & \twoheadrightarrow & A_L^T
\end{array}$$

Here, the middle row is sequence (2.11) before tensoring with $\mathbb{Z}_p$. We obviously have an equality

$$i_G(\operatorname{cok}\phi_S^T \otimes \mathbb{Z}_p) = i_G(\operatorname{cok}\psi \otimes \mathbb{Z}_p) - i_G(A_L^T \otimes \mathbb{Z}_p) \tag{2.21}$$

in $K_0(\mathbb{Z}_pG_-, \mathbb{Q}_p)$.

Lemma 2.3.6 *The element $i_G(\operatorname{cok}\psi \otimes \mathbb{Z}[\frac{1}{2}]) \in K_0(\mathbb{Z}[\frac{1}{2}]G_-, \mathbb{Q})$ has representing homomorphism*

$$\chi \mapsto \frac{R_{\phi_S}(\check{\chi})}{\prod_{\mathfrak{p}\in S(K)}(-\log N(\mathfrak{p}))^{\dim V_\chi}},$$

where $S(K) := \{\mathfrak{P} \cap K | \mathfrak{P} \in S\}$.

PROOF. Let us denote the inclusion $E_S^{T,-} \rightarrowtail (\mathbb{Z}G_-)^{s^*}$ by μ. Define a map

$$\begin{array}{rcl}
\operatorname{Log} : (\mathbb{Z}G_-)^{s^*} & \longrightarrow & \mathbb{R} \otimes (\mathbb{Z}G_-)^{s^*} \\
(x_1, \dots, x_{s^*}) & \mapsto & (-\log N(\mathfrak{p}_1) \otimes x_1, \dots, -\log N(\mathfrak{p}_{s^*}) \otimes x_{s^*}),
\end{array}$$

where we have numbered the primes in $S(K) = \{\mathfrak{p}_1, \dots, \mathfrak{p}_{s^*}\}$. Then

$$\psi = \mu \circ \phi_S \circ \rho_S^{-1},$$

$$\lambda_S^- = \rho_S^{-1} \circ \operatorname{Log} \circ \mu,$$

where λ_S^- is the restriction of the Dirichlet map to minus parts. Hence, $\lambda_S^- \phi_S = \rho_S^{-1} \circ \mathrm{Log} \circ \psi \circ \rho_S$, and $i_G(\operatorname{cok} \psi \otimes \mathbb{Z}[\frac{1}{2}])$ has representing homomorphism

$$\begin{aligned} \chi \quad \mapsto \quad & \det(\psi | \mathrm{Hom}_{\mathbb{C}G}(V_\chi, (\mathbb{C}G_-)^{s^*})) \\ &= \frac{R_{\phi_S}(\check{\chi})}{\det(\mathrm{Log} | \mathrm{Hom}_{\mathbb{C}G}(V_\chi, (\mathbb{C}G_-)^{s^*}))} \\ &= \frac{R_{\phi_S}(\check{\chi})}{\prod_{\mathfrak{p} \in S(K)} (-\log N(\mathfrak{p}))^{\dim V_\chi}}. \end{aligned}$$

This completes the proof. □

Note that the Stark-Tate regulator occurring in the representing homomorphism of $i_G(\operatorname{cok} \psi \otimes \mathbb{Z}[\frac{1}{2}])$ is closely related to the modified Stark-Tate regulator; more precisely, we have (cf. the proof of Theorem 1.5.8)

$$\frac{R_\phi^{\mathrm{mod}}(\chi)}{R_{\phi_S}(\chi)} = \prod_{\mathfrak{P} \in S^*_{\mathrm{ram}}} \left(- \frac{\log N(\mathfrak{P})}{|G_\mathfrak{P}|} \right)^{\dim V_\chi^{G_\mathfrak{P}}} .$$

If we now combine the equations (2.19) and (2.21) with the above Lemma and Proposition 2.3.5, we get Theorem 2.3.1 by an easy computation. □

Chapter 3

Iwasawa theory

As an application of Theorem 2.3.1 we are going to prove the minus part of the LRNC at a prime $p \neq 2$ if L/K is an abelian CM-extension fulfilling the assumptions of the theorem; actually, we need to work under a slightly more restrictive hypothesis on the primes above p. We additionally require the vanishing of the μ-invariant of the standard Iwasawa module (all this will be made explicit below). But we will see in the appendix how to remove this assumption for some special cases, including the case $p \nmid |G|$. The main ingredient of the proof turns out to be the validity of the Iwasawa main conjecture for abelian extensions.

3.1 Passing to the limit

Let L/K be an abelian CM-extension with Galois group G and $p \neq 2$ a finite rational prime such that all primes $\mathfrak{p} \subset K$ above p are tamely ramified in L/K or $j \in G_{\mathfrak{p}}$. Here, we write $G_{\mathfrak{p}}$ instead of $G_{\mathfrak{P}}$, since the decomposition group only depends on the prime $\mathfrak{p}$ in K if G is abelian. We will accordingly write $I_{\mathfrak{p}}$, $\phi_{\mathfrak{p}}$ etc. As it is required for the use of Theorem 2.3.1, we choose a finite prime $\mathfrak{P}_0$ of L such that $1 - \zeta \notin \prod_{g \in G/G_{\mathfrak{P}_0}} \mathfrak{P}_0^g$ for all roots of unity $\zeta \neq 1$ in L. We may assume that $\mathfrak{P}_0$ is unramified in L/K and does not divide p. Indeed, it would suffice to ask for a corresponding condition on $\mathfrak{P}_0$ for all p-power roots of unity in L, since we tensor with $\mathbb{Z}_p$. Hence, any prime which lies not above p will do.
As before we define a finite set of places of L

$$T = S_{\text{ram}} \setminus (S_{\text{ram}} \cap S_p) \cup \{\mathfrak{P}_0^g | g \in G\}\,, \tag{3.1}$$

and set $A_L^T = \text{cl}_L^{T,-}$. Then $A_L^T \otimes \mathbb{Z}_p$ is c.t. by Theorem 2.1.1.
Let L_∞ and K_∞ be the cyclotomic $\mathbb{Z}_p$-extensions of L and K, respectively. We denote the Galois group of K_∞/K by Γ_K. Hence, Γ_K is isomorphic to $\mathbb{Z}_p$, and we fix a topological generator γ_K. Furthermore, we denote the n-th layer in the cyclotomic extension K_∞/K by K_n such that K_n/K is cyclic of

order p^n. Accordingly, we set $\Gamma_L = \mathrm{Gal}(L_\infty/L)$ with a topological generator γ_L whose restriction to K_∞ is γ_K^a for an appropriate integer a. We enumerate the intermediate fields starting with $L = L_a$ such that L_n/L is cyclic of order p^{n-a}. This is because then L_n is the smallest intermediate field of L_∞/L which lies above K_n. It may also be convenient to define $L_n = L$ if $n \leq a$. Let

$$T_n := \{\mathfrak{P}_n \subset L_n | \mathfrak{P}_n \cap L \in T\},$$

so $T_0 = T$ and $A_{L_n}^{T_n} \otimes \mathbb{Z}_p$ is $\mathrm{Gal}(L_n/K_n)$-c.t., since each of the extensions L_n/K_n inherits the required properties from the extension L/K. We define

$$X_T^- := \lim_{\leftarrow} A_{L_n}^{T_n} \otimes \mathbb{Z}_p.$$

We denote the Galois group of L_∞/K by $\mathcal{G}$, hence

$$\mathcal{G} = \tilde{G} \times \Gamma_K,$$

where $\tilde{G}$ is a subgroup of G. Then the completed group ring $\mathbb{Z}_p[[\mathcal{G}]]$ is isomorphic to $\Lambda[\tilde{G}]$, where Λ is the Iwasawa algebra $\mathbb{Z}_p[[T]]$. Since we are going to use some of the results in [Gr2], we set $\gamma_K = 1 - T$ as in loc.cit.
There is an exact sequence of type (2.8) for each layer n. In the limit this yields an exact sequence (cf. [Gr2], Proposition 6)

$$\mathbb{Z}_p(1) \rightarrowtail \bigoplus_{\mathfrak{p} \in T(K)} Z_{\mathfrak{p}}(1)^- \to X_T^- \twoheadrightarrow X_{\mathrm{std}}^- \tag{3.2}$$

if $\zeta_p \in L$, and without the $\mathbb{Z}_p(1)$ term if $\zeta_p \notin L$. Here, X_{std} is the standard Iwasawa module which is the projective limit of the p-parts of the class groups in the cyclotomic tower over L, and $Z_{\mathfrak{p}}(1)$ is the first Tate twist of

$$Z_{\mathfrak{p}} = \mathrm{ind}_{\mathcal{G}_{\mathfrak{p}}}^{\mathcal{G}} \mathbb{Z}_p = \mathbb{Z}_p[[\Gamma_K \times \tilde{G}/\tilde{I}_{\mathfrak{p}}]]/(1 - \phi_{\mathfrak{p}}),$$

where we now write $\phi_{\mathfrak{p}}$ for the Frobenius automorphism at $\mathfrak{p}$ in the Galois group $\mathcal{G}$. The basic facts about the Iwasawa module X_T^- are summarized in the following Proposition.

Proposition 3.1.1 *The Iwasawa module X_T^- is a finitely generated, torsion $\mathbb{Z}_p[[\mathcal{G}]]_-$-module, which has no non-trivial finite submodules and*

$$\mathrm{pd}_{\mathbb{Z}_p[[\mathcal{G}]]_-}(X_T^-) \leq 1.$$

PROOF. This is Proposition 7 in [Gr2], where the ramification above p is assumed to be tame. But what is needed is just the cohomological triviality of the ray class groups $A_{L_n}^{T_n} \otimes \mathbb{Z}_p$. □

The Fitting ideal of X_T^- is described in terms of p-adic L-functions. To make this explicit we have to introduce some further notation. Let $\kappa : \mathcal{G} \to \mathbb{Z}_p^\times$ denote the cyclotomic character and define $u = \kappa(\gamma_K)$. Any character ψ of $\mathcal{G}$

with open kernel can be written as $\psi = \chi \otimes \rho$, where χ is a character of $\tilde{G}$ and ρ is trivial on $\tilde{G}$ (so χ is of type S and ρ is of type W in the terminology of [Wi1]). If χ is an odd character and S a set of places of K containing all the primes above p, there exists a well-defined element $f_{\chi,S}(T) \in Quot(\mathbb{Z}_p(\chi)[[T]])$ determined by

$$f_{\chi,S}(u^s - 1) = L_{p,S}(s, \omega\chi^{-1}),\ \ s = 1, 2, 3, \dots$$

where ω is the Teichmüller character[1] on $L(\zeta_p)/K$. This definition of $f_{\chi,S}$ follows the convention of Washington's book [Wa], and is used in [Gr2]. It is also usual to replace the argument s on the right hand side by $1 - s$, but this makes no essential difference.
For all χ of type S and ρ of type W we have (cf. [Gr2], Lemma 7)

$$f_{\chi\otimes\rho,S}(T) = f_{\chi,S}(\rho(\gamma_K)(1 + T) - 1). \tag{3.3}$$

For this, note that in the notation of [Wi1] we have an equality

$$f_{\chi\otimes\rho,S}(T) = \frac{G_{\omega\chi^{-1}\otimes\rho,S}(u(1+T)^{-1} - 1)}{H_{\omega\chi^{-1}\otimes\rho,S}(u(1+T)^{-1} - 1)}$$

and a similar formula holds for the right hand side. The Iwasawa series $f_{\chi\otimes\rho,S}(T)$ glue together for varying characters, i.e. there exists a unique element $\Phi_S \in Quot(\mathbb{Z}_p[[\mathcal{G}]])^-$ such that for all odd characters $\psi = \chi \otimes \rho$ of $\mathcal{G}$ we have (cf. [Gr2], Proposition 11)

$$\psi(\Phi_S) = f_{\chi,S}(\rho(\gamma_K) - 1).$$

Let $\mathfrak{p} \nmid p$ be a finite prime of K. Put

$$\xi_{\mathfrak{p}} = \frac{\kappa(\phi_{\mathfrak{p}}) - \phi_{\mathfrak{p}}}{1 - \phi_{\mathfrak{p}}}\,\varepsilon_{\mathfrak{p}} + 1 - \varepsilon_{\mathfrak{p}} \in Quot(\mathbb{Z}_p[[\mathcal{G}]]), \tag{3.4}$$

where $\varepsilon_{\mathfrak{p}} = |I_{\mathfrak{p}}|^{-1} N_{I_{\mathfrak{p}}} \in \mathbb{Q}_p\tilde{G} \subset \mathbb{Q}_p[[\mathcal{G}]]$. If T is a finite set of primes of L which contains no prime above p, define

$$\Psi_T = \left(\prod_{\mathfrak{p}\in T(K)} \xi_{\mathfrak{p}}\right) \cdot \Phi_{T(K)\cup S_p}.$$

If T is the set of places defined in (3.1), we have (cf. [Gr2], Proposition 9)

$$\frac{1 - j}{2}\Psi_T \in \mathbb{Z}_p[[\mathcal{G}]]^-.$$

The Iwasawa main conjecture is the main ingredient in proving

[1]Do not confuse with the group ring element ω occurring in Proposition 2.2.2. ω will always denote the Teichmüller character in what follows.

Theorem 3.1.2 *Let T be the set of places of L defined in (3.1) and μ_- the μ-invariant of the standard Iwasawa module X^-_{std}. Then it holds:*

(1) The Fitting ideal of $\mathbb{Q}_p X_T^-$ is generated by Ψ_T.

(2) If $\mu_- = 0$, we actually have

$$\mathrm{Fitt}_{\mathbb{Z}_p[[\mathcal{G}]]_-}(X_T^-) = (\Psi_T).$$

PROOF. If the ramification above p is almost tame, this is Proposition 8 and Theorem 6 in [Gr2]. But once more the condition on the ramification is only needed to guarantee the cohomological triviality of $A_L^T \otimes \mathbb{Z}_p$. □

REMARK. If we denote the total ring of fractions of $\mathbb{Z}_p[[\mathcal{G}]]_-$ by $\mathcal{Q}(\mathbb{Z}_p[[\mathcal{G}]]_-)$, there is the Localization Sequence (cf. (1.3))

$$K_1(\mathbb{Z}_p[[\mathcal{G}]]_-) \to K_1(\mathcal{Q}(\mathbb{Z}_p[[\mathcal{G}]]_-)) \xrightarrow{\partial} K_0T(\mathbb{Z}_p[[\mathcal{G}]]_-) \to K_0(\mathbb{Z}_p[[\mathcal{G}]]_-).$$

Since the determinant yields an isomorphism

$$K_1(\mathcal{Q}(\mathbb{Z}_p[[\mathcal{G}]]_-)) \simeq (\mathcal{Q}(\mathbb{Z}_p[[\mathcal{G}]]_-))^\times,$$

we can view $\frac{1-j}{2}\Psi_T$ as an element of $K_1(\mathcal{Q}(\mathbb{Z}_p[[\mathcal{G}]]_-))$. So (2) of the above theorem means that $\mu_- = 0$ implies $\partial(\frac{1-j}{2}\Psi_T) = [X_T^-]$. Indeed, one should think of the claim in (2) as a reformulation of the equivariant Iwasawa main conjecture (for the case at hand) which is known to be true if $\mathcal{G}$ is abelian and $\mu = 0$ by Theorem 11 in [RW3].

Lemma 3.1.3 *Let ψ be a character of $\mathcal{G}$ with open kernel and S a set of places of K that contains all the p-adic places. Put*

$$S_\psi = \{\mathfrak{p} \in S | I_\mathfrak{p} \not\subset \ker(\psi)\} \cup S_p$$

and write the Frobenius automorphism at a prime $\mathfrak{p}$ as $\phi_\mathfrak{p} = \sigma_\mathfrak{p}\gamma_K^{c_\mathfrak{p}}$, where $\sigma_\mathfrak{p} \in \tilde{G}$ and $c_\mathfrak{p} \in \mathbb{Z}_p$.

(1) Let χ be a character of $\tilde{G}$. Then

$$L_{p,S}(s, \omega\chi^{-1}) = L_{p,S_\chi}(s, \omega\chi^{-1}) \prod_{\mathfrak{p}\in S\setminus S_\chi} (1 - \chi^{-1}(\sigma_\mathfrak{p})u^{-s\cdot c_\mathfrak{p}}).$$

(2) We have an equality

$$f_{\psi,S}(T) = f_{\psi,S_\psi}(T) \prod_{\mathfrak{p}\in S\setminus S_\psi} (1 - \psi^{-1}(\phi_\mathfrak{p})(1+T)^{-c_\mathfrak{p}}).$$

PROOF. (1) is well known and follows by evaluating both sides of the equation at $s = 1 - n$, where $n \equiv 0 \bmod (p-1)$. (2) is an easy consequence of (1) using formula (3.3) for the character $\psi = \chi \otimes \rho$ with a $\tilde{G}$-character χ. □

The following corollary will be important in the sequel.

Corollary 3.1.4 *Let T be the set of places of L defined in (3.1) and S_1 be the set of places of L which are wildly ramified in L/K. Each character χ of G can be viewed as a character of $\mathcal{G}$ and, if χ is odd,*

$$\chi(\Psi_T) = \chi(\theta_{S_1}^T) \cdot \prod_{\mathfrak{p} \in S_p \cap S_{\mathrm{tram}}} (1 - \chi(\varepsilon_{\mathfrak{p}} \phi_{\mathfrak{p}}^{-1})),$$

where the product runs over all p-adic places of K which are at most tamely ramified.

PROOF. Write $\chi = \chi' \otimes \rho$, where χ' is a character of $\tilde{G}$ and ρ is of type W. Since only p-adic primes ramify in the cyclotomic towers over K and L, we have $\Sigma_\chi = \Sigma_{\chi'}$, where $\Sigma = T(K) \cup S_p$. At first, we determine $\chi'(\Psi_T) \in \mathbb{Z}_p(\chi')[[T]]$. With the notation of Lemma 3.1.3 we have

$$\begin{aligned}
\chi'(\Psi_T) &= \prod_{\mathfrak{p} \in T(K)} \frac{\kappa(\phi_{\mathfrak{p}}) - \chi'(\sigma_{\mathfrak{p}})\gamma_K^{c_{\mathfrak{p}}}}{1 - \chi'(\sigma_{\mathfrak{p}})\gamma_K^{c_{\mathfrak{p}}}} f_{\chi',\Sigma}(-T) \\
&\overset{(*)}{=} \left(\prod_{\mathfrak{p} \in T(K)} \frac{\kappa(\phi_{\mathfrak{p}}) - \chi'(\sigma_{\mathfrak{p}})\gamma_K^{c_{\mathfrak{p}}}}{1 - \chi'(\sigma_{\mathfrak{p}})\gamma_K^{c_{\mathfrak{p}}}} (1 - \chi'(\sigma_{\mathfrak{p}})^{-1}\gamma_K^{-c_{\mathfrak{p}}}) \right) f_{\chi',\Sigma_{\chi'}}(-T) \\
&= \prod_{\mathfrak{p} \in T(K)} (1 - \chi'(\sigma_{\mathfrak{p}})^{-1}\gamma_K^{-c_{\mathfrak{p}}}\kappa(\phi_{\mathfrak{p}})) f_{\chi',\Sigma_{\chi'}}(-T),
\end{aligned}$$

where (*) holds by means of (2) of Lemma 3.1.3. Since

$$\rho(f_{\chi',\Sigma_{\chi'}}(-T)) = f_{\chi',\Sigma_{\chi'}}(\rho(\gamma_K) - 1) = f_{\chi,\Sigma_\chi}(0) = L_{S_\chi}(0, \chi^{-1}),$$

we get

$$\begin{aligned}
\chi(\Psi_T) &= \rho(\chi'(\Psi_T)) \\
&= \prod_{\mathfrak{p} \in T(K)} (1 - \chi(\phi_{\mathfrak{p}})^{-1}\kappa(\phi_{\mathfrak{p}})) L_{S_\chi}(0, \chi^{-1}) \\
&= \prod_{\mathfrak{p} \in T(K)} (1 - \chi(\phi_{\mathfrak{p}})^{-1} q_{\mathfrak{p}}) \prod_{\mathfrak{p} \in S_p} (1 - \chi(\varepsilon_{\mathfrak{p}}\phi_{\mathfrak{p}}^{-1})) L_{S_\infty}(0, \chi^{-1}) \\
&= \chi(\theta_{S_1}^T) \cdot \prod_{\mathfrak{p} \in S_p \cap S_{\mathrm{tram}}} (1 - \chi(\varepsilon_{\mathfrak{p}}\phi_{\mathfrak{p}}^{-1})),
\end{aligned}$$

where as before $q_{\mathfrak{p}} = N(\mathfrak{p})$. □

3.2 The descent

We are going to use an idea, which originates from [Wi2], in the extended version of [Gr1], where the author proves Brumer's conjecture for a special class of CM-extensions. Note that the class of CM-extensions treated here includes the class of loc. cit. The same approach is also used in [Ku] to compute the Fitting ideals of minus class groups of absolute abelian CM-fields. But before we go for this, we look at a special case, where a rather restrictive condition forces the Euler factors at p to become units in $\mathbb{Z}_pG_-$.

Proposition 3.2.1 *Let L/K be an abelian CM-extension with Galois group G and $p \neq 2$ a rational prime. Let T be the set of places of L defined in (3.1) and S_1 be the set of all wildly ramified primes. Suppose that $\mu_- = 0$ and $j \in G_\mathfrak{p}$ for all primes $\mathfrak{p}$ of K above p.*
Then $\theta^T_{S_1}$ generates the Fitting ideal $\mathrm{Fitt}_{\mathbb{Z}_pG_-}(A^T_L \otimes \mathbb{Z}_p)$. In particular, the minus part of the LRNC at p is true.

PROOF. The canonical restriction map $X^-_T \to A^T_L \otimes \mathbb{Z}_p$ is an epimorphism, since the cokernel is a quotient of Γ_L which has trivial j-action. By general properties of Fitting ideals we have

$$\mathrm{Fitt}_{\mathbb{Z}_pG_-}(X^-_T/\gamma_L - 1) \subset \mathrm{Fitt}_{\mathbb{Z}_pG_-}(A^T_L \otimes \mathbb{Z}_p),$$

and the Fitting ideal on the left hand side is generated by $\Psi_T \bmod (\gamma_L - 1)$ by Theorem 3.1.2. Corollary 3.1.4 now implies that

$$\Psi_T \bmod (\gamma_L - 1) = \theta^T_{S_1} \prod_{\mathfrak{p} \in S_p \cap S_{\mathrm{tram}}} (1 - \varepsilon_\mathfrak{p} \phi_\mathfrak{p}^{-1}).$$

But the product on the right hand side is a unit in $\mathbb{Z}_pG_-$, since $j \in G_\mathfrak{p}$ for these primes. Hence $\theta^T_{S_1} \in \mathrm{Fitt}_{\mathbb{Z}_pG_-}(A^T_L \otimes \mathbb{Z}_p)$. Finally, Proposition 2.2.5 and 2.2.7 imply that $\theta^T_{S_1}$ has to be a generator of the Fitting ideal.
The minus part of the LRNC at p now follows from Theorem 2.3.1. □

Now we use the method in [Gr1] to prove the minus part of the LRNC at p without the additional assumption of Proposition 3.2.1. But this works only for primes p such that $L^{\mathrm{cl}} \not\subset (L^{\mathrm{cl}})^+(\zeta_p)$, where L^{cl} denotes the normal closure of L over $\mathbb{Q}$, which is again a CM-field. This condition particularly forces $\zeta_p \notin L$. But note that this condition holds for almost all primes p, since each prime for which it fails has to ramify in $L^{\mathrm{cl}}/\mathbb{Q}$. Our main result is

Theorem 3.2.2 *Let L/K be an abelian CM-extension with Galois group G and $p \neq 2$ a rational prime. Let T be the set of places of L defined in (3.1) and S_1 be the set of all wildly ramified primes. Suppose that $\mu_- = 0$ and that each prime $\mathfrak{p}$ above p ramifies at most tame or $j \in G_\mathfrak{p}$. Moreover, assume that $j \in G_\mathfrak{p}$ for all primes $\mathfrak{p}$ of K above p whenever $L^{\mathrm{cl}} \subset (L^{\mathrm{cl}})^+(\zeta_p)$.*
Then $\theta^T_{S_1}$ generates the Fitting ideal $\mathrm{Fitt}_{\mathbb{Z}_pG_-}(A^T_L \otimes \mathbb{Z}_p)$. In particular, the minus part of the LRNC at p is true.

REMARK. The vanishing of μ_- is only required for computing the Fitting ideal of X_T^- (cf. Theorem 3.1.2). As already mentioned, we will show in the appendix that we can remove this hypothesis for some special cases, including the case $p \nmid |G|$.

PROOF. The assertion follows from Proposition 3.2.1 if $L^{\mathrm{cl}} \subset (L^{\mathrm{cl}})^+(\zeta_p)$. Hence, we may assume that this is not the case in the following.
We state the following result, which is Proposition 4.1 in [Gr1].

Proposition 3.2.3 *Let p be a prime such that $L^{\mathrm{cl}} \not\subset (L^{\mathrm{cl}})^+(\zeta_p)$ and $N \in \mathbb{N}$. Then there exist infinitely many primes r such that*

- $r \equiv 1 \bmod p^N$
- *$j \in G_{\mathfrak{r}}$ for each prime $\mathfrak{r}$ in K above r*
- *the Frobenius automorphism at p in the extension $\mathbb{Q}(\zeta_r)/\mathbb{Q}$ generates $\mathrm{Gal}(E/\mathbb{Q})$, where E is the subfield of $\mathbb{Q}(\zeta_r)$ such that $[E:\mathbb{Q}] = p^N$.*

Let N be a large integer to be chosen later, and choose a prime r as in the Proposition which does not ramify in $L^{\mathrm{cl}}/\mathbb{Q}$. The extension $E/\mathbb{Q}$ is cyclic of degree p^N, and we denote the corresponding Galois group by C_N. It is generated by the Frobenius automorphism $\mathrm{Frob}_p \in C_N$. Let $L' = LE$ and $K' = KE$. Then L'/K is an abelian extension with Galois group $G' = G \times C_N$, and the only new ramification occurs above r. Moreover, the primes $\mathfrak{r}$ above r satisfy both of our standard conditions: They are tamely ramified and $j \in G_{\mathfrak{r}}$. Set $T' = \{\mathfrak{P}' \subset L' : \mathfrak{P}' \cap L \in T\}$ and $T_0' = T' \cup \{\mathfrak{R}' \in L' : \mathfrak{R}' \mid r\}$. There is an exact sequence

$$\left(\mathfrak{o}_{L'}/\prod_{\mathfrak{R}'|r}\mathfrak{R}'\right)^{\times,-} \otimes \mathbb{Z}_p \rightarrowtail A_{L'}^{T_0'} \otimes \mathbb{Z}_p \twoheadrightarrow A_{L'}^{T'} \otimes \mathbb{Z}_p.$$

We claim that the leftmost term is trivial, and hence $A_{L'}^{T'} \otimes \mathbb{Z}_p \simeq A_{L'}^{T_0'} \otimes \mathbb{Z}_p$ is c.t. by Theorem 2.1.1. To see this let $\mathfrak{r}$ be a prime in K above r, and $\mathfrak{R}'$ a prime in L' above $\mathfrak{r}$. Since $j \in G_{\mathfrak{r}}$, it acts on the corresponding residue field extension of degree $f_{\mathfrak{r}}$, say. Therefore, $(\mathfrak{o}_{L'}/\mathfrak{R}')^{\times,-}$ has exactly $q_{\mathfrak{r}}^{f_{\mathfrak{r}}/2} + 1$ elements, where $q_{\mathfrak{r}} = N(\mathfrak{r})$ is a power of r. But thanks to the first condition on r we have $q_{\mathfrak{r}}^{f_{\mathfrak{r}}/2} + 1 \equiv 2 \not\equiv 0 \bmod p$. Hence, the leftmost term vanishes, since we are only dealing with p-parts.

For the same reasons as in Proposition 3.2.1 the natural restriction map $A_{L'}^{T'} \otimes \mathbb{Z}_p \twoheadrightarrow A_L^T \otimes \mathbb{Z}_p$ is surjective. The composite map

$$A_{L'}^{T'} \otimes \mathbb{Z}_p \xrightarrow{\mathrm{res}} A_L^T \otimes \mathbb{Z}_p \rightarrow A_{L'}^{T'} \otimes \mathbb{Z}_p$$

is given by the norm N_{C_N}, and the kernel of the norm is just $\Delta C_N \cdot A_{L'}^{T'} \otimes \mathbb{Z}_p$. Therefore, the restriction map induces an isomorphism

$$(A_{L'}^{T'} \otimes \mathbb{Z}_p)_{C_N} \xrightarrow{\simeq} A_L^T \otimes \mathbb{Z}_p. \qquad (3.5)$$

As before, we build the cyclotomic tower over L' and set $\Gamma_{L'} = \mathrm{Gal}(L'_\infty/L')$ and $\mathcal{G}' = \mathrm{Gal}(L'_\infty/K) = \mathcal{G} \times C_N$. We define the projective limit of the ray class groups $A_{L'_n}^{T'_n} \otimes \mathbb{Z}_p$ to be $X_{T'}^-$, which is a finitely generated, torsion $\mathbb{Z}_p[[\mathcal{G}']]_-$-module of projective dimension at most 1 by Proposition 3.1.1. Since we assume $\mu_- = 0$ for the cyclotomic $\mathbb{Z}_p$-extension L_∞/L, the same holds for the cyclotomic $\mathbb{Z}_p$-extension over L' by Theorem 11.3.8 in [NSW]. Theorem 3.1.2 implies

$$\mathrm{Fitt}_{\mathbb{Z}_p[[\mathcal{G}']]_-}(X_{T'}^-) = (\Psi_{T'_0}).$$

Set $u_p = \prod_{\mathfrak{p} \in S_p \cap S_{\mathrm{tram}}} (1 - \varepsilon_\mathfrak{p} \phi_\mathfrak{p}^{-1}|_{L'}) \in \mathbb{Z}_p G'$. As in Proposition 3.2.1, the canonical restriction map $X_{T'}^- \to A_{L'}^{T'} \otimes \mathbb{Z}_p$ is an epimorphism, and therefore the Fitting ideal of $A_{L'}^{T'} \otimes \mathbb{Z}_p$ contains $u_p \cdot \theta_{S'_1}^{T'_0}$ using Corollary 3.1.4. Here, S'_1 are the primes in L' above those in S_1, i.e. S'_1 contains all the wildly ramified primes of the extension L'/K.
Let M be a natural number, $M \leq N$, and $\nu = \sum_{i=0}^{p^M-1} \mathrm{Frob}_p^{ip^{N-M}} \in \mathbb{Z}_p C_N$.

Lemma 3.2.4 *Let f be the least common multiple of the residual degrees $f_\mathfrak{p}$ of all $\mathfrak{p} \in S_p$ corresponding to the extension $K/\mathbb{Q}$. If $N - M \geq v_p(|G| \cdot f)$, then the element u_p is a nonzerodivisor in $\mathbb{Z}_p G'/\nu \mathbb{Z}_p G'$.*

PROOF. The proof of Proposition 4.6 in [Gr1] carries over to the present situation. □

Corollary 3.2.5 *Under the same hypothesis concerning ν as in Lemma 3.2.4 we have:*

(1) $u_p \cdot \theta_{S'_1}^{T'_0}$ is a nonzerodivisor in $R' := \mathbb{Z}_p G'_- / \nu \mathbb{Z}_p G'_-$.

(2) $(X_{T'}^-)_{\Gamma_{L'}}/\nu$ has projective dimension at most 1 over R', and its Fitting ideal is generated by $u_p \cdot \theta_{S'_1}^{T'_0} \bmod \nu$.

PROOF. Again the proof of Corollary 4.7 in [Gr1] remains unchanged. But note that (1) is clear by Lemma 3.2.4, since ν is a zerodivisor in $\mathbb{Z}_p G'_-$, but $\theta_{S'_1}^{T'_0}$ is not. □

We claim that there is an exact sequence

$$\bigoplus_{\mathfrak{p} \in S_p} \mathbb{Z}_p[G'/G'_\mathfrak{p}]^- \to (X_{T'}^-)_{\Gamma_{L'}} \twoheadrightarrow A_{L'}^{T'} \otimes \mathbb{Z}_p \tag{3.6}$$

of $\mathbb{Z}_p G'$-modules, where we can replace the set S_p by $S_p \cap S_{\mathrm{tram}}$, since $\mathbb{Z}_p[G'/G'_\mathfrak{p}]^-$ vanishes if $j \in G'_\mathfrak{p}$. Note that an analogous sequence is well known if we replace the ray class groups by ordinary class groups (see [Gr1], p. 530 or [Wa]). We postpone the proof and first continue with the proof of Theorem 3.2.2. We need the following result about Fitting ideals (cf. Lemma 7.1 in [Ku]).

Lemma 3.2.6 *Let R be a commutative ring and $M_1 \to M_2 \twoheadrightarrow M_3$ an exact sequence of R-modules. Then*

$$\mathrm{Fitt}_R(M_1)\mathrm{Fitt}_R(M_3) \subset \mathrm{Fitt}_R(M_2).$$

If we tensor the exact sequence (3.6) with R' and apply the above Lemma, we get

$$\mathrm{Fitt}_{R'}(A_{L'}^{T'} \otimes \mathbb{Z}_p/\nu) \cdot \mathrm{Fitt}_{R'}\Big(\bigoplus_{\mathfrak{p} \in S_p \cap S_{\mathrm{tram}}} R'/(1 - \varepsilon_{\mathfrak{p}}\phi_{\mathfrak{p}}^{-1}|_{L'})\Big) \subset \mathrm{Fitt}_{R'}((X_{T'})_{\Gamma_{L'}}/\nu).$$

Hence, $\mathrm{Fitt}_{R'}(A_{L'}^{T'} \otimes \mathbb{Z}_p/\nu) \subset (\theta_{S_1'}^{T_0'} \bmod \nu)$ by Corollary 3.2.5. The augmentation map $\mathrm{aug}_G^{G'} : \mathbb{Z}_pG' \twoheadrightarrow \mathbb{Z}_pG$ induces the first isomorphism in

$$(A_{L'}^{T'} \otimes \mathbb{Z}_p/\nu) \otimes \mathbb{Z}_pG_- \simeq (A_{L'}^{T'} \otimes \mathbb{Z}_p)_{C_N}/\mathrm{aug}_G^{G'}(\nu) \simeq A_L^T \otimes \mathbb{Z}_p/p^M,$$

whereas the second isomorphism derives from (3.5). Since the Fitting ideal behaves well under base change, we get

$$\mathrm{Fitt}_{\mathbb{Z}_pG_-/p^M}(A_L^T \otimes \mathbb{Z}_p/p^M) \subset (\mathrm{aug}_G^{G'}(\theta_{S_1'}^{T_0'}) \bmod p^M).$$

But $\mathrm{aug}_G^{G'}(\theta_{S_1'}^{T_0'}) = \prod_{\mathfrak{r} \in S_r}(1 - \phi_{\mathfrak{r}}^{-1}|_L q_{\mathfrak{r}}) \cdot \theta_{S_1}^T$ and the product over the primes above r is a unit in $\mathbb{Z}_pG_-$. Therefore

$$\mathrm{Fitt}_{\mathbb{Z}_pG_-}(A_L^T \otimes \mathbb{Z}_p) \subset (\theta_{S_1}^T) + p^M \cdot \mathbb{Z}_pG_-,$$

and since we can choose M arbitrarily large, we actually get

$$\mathrm{Fitt}_{\mathbb{Z}_pG_-}(A_L^T \otimes \mathbb{Z}_p) \subset (\theta_{S_1}^T).$$

As in the proof of Proposition 3.2.1, $\theta_{S_1}^T$ now has to be a generator of the Fitting ideal by Proposition 2.2.5 and 2.2.7.
The minus part of the LRNC at p again follows from Theorem 2.3.1. □

We are left with the existence of sequence (3.6). Indeed, we prove a more general result.

Proposition 3.2.7 *Let L/K be a Galois CM-extension with Galois group G, $p \neq 2$ a rational prime and T a finite G-invariant set of places of L such that $T \cap S_p = \emptyset$. If X_T^- denotes the projective limit of the ray class groups $A_{L_n}^{T_n} \otimes \mathbb{Z}_p$, where T_n consists of all primes in the n-th layer L_n of the cyclotomic $\mathbb{Z}_p$-extension above the primes in T, there is an exact sequence of $\mathbb{Z}_pG_-$-modules*

$$\bigoplus_{\mathfrak{P} \in S_p^*} \mathbb{Z}_p[G/G_{\mathfrak{P}}]^- \to (X_T^-)_{\Gamma_L} \twoheadrightarrow A_L^T \otimes \mathbb{Z}_p.$$

PROOF. For the same reasons as in Proposition 3.2.1, the canonical restriction map $X_T^- \to A_L^T \otimes \mathbb{Z}_p$ is an epimorphism and clearly factors through $(X_T^-)_{\Gamma_L}$.
By class field theory, each ray class group $\mathrm{cl}_{L_n}^{T_n} \otimes \mathbb{Z}_p$ is the Galois group of a finite abelian p-extension M_n/L_n. Then the projective limit X_T of these ray class groups is the Galois group of the extension M_∞/L_∞, where $M_\infty = \bigcup_{n \in \mathbb{N}} M_n$. We put $\mathcal{X} = \mathrm{Gal}(M_\infty/L)$.
Let $\mathfrak{P}_1, \dots, \mathfrak{P}_s$ be the primes in L above p. Exactly these primes ramify in L_∞/L, and we denote the finitely many primes in L_∞, which lie above $\mathfrak{P}_1, \dots, \mathfrak{P}_s$, by $\mathfrak{P}_{ik}^\infty$, $1 \le i \le s$. Moreover, we choose above each $\mathfrak{P}_{ik}^\infty$ a prime $\tilde{\mathfrak{P}}_{ik}$ in M_∞, and denote its inertia group in M_∞/L by $I_{ik} \le \mathcal{X}$.
We obviously have an isomorphism $\mathcal{X}/X_T \simeq \Gamma_L$. So we can pick a preimage $\gamma \in \mathcal{X}$ of γ_L, and thus

$$\mathcal{X} = X_T \cdot \overline{\langle \gamma \rangle}. \tag{3.7}$$

The elements in G act on $\mathcal{X}$ via group conjugation, and we may assume that $\gamma^j = \gamma$ by replacing γ by $\gamma^{(1-j)/2}$. The condition on the set T forces that the extension M_∞/L_∞ does not ramify above p. Therefore $I_{ik} \cap X_T = 1$, and we get inclusions

$$I_{ik} \rightarrowtail \mathcal{X}/X_T = \Gamma_L.$$

Hence, each I_{ik} is isomorphic to $\Gamma_L^{p^{n_{ik}}}$ for an appropriate integer n_{ik}. We fix a topological generator σ_{ik} of I_{ik} which maps to $\gamma_L^{p^{n_{ik}}}$ via the above inclusion. But for fixed i, each two of these inertia groups are conjugate to each other, and hence $n_{ik} = n_{i1} =: n_i$ for all k. Corresponding to (3.7) we write $\sigma_{ik} = a_{ik}\gamma^{p^{n_i}}$ with $a_{ik} \in X_T$.
Because of the obvious exact sequence

$$\mathrm{Gal}(M_\infty/M_0) \rightarrowtail \mathcal{X} \twoheadrightarrow \mathrm{cl}_L^T \otimes \mathbb{Z}_p$$

we are interested in the Galois group $\mathrm{Gal}(M_\infty/M_0)$. We claim that it equals the subgroup $\mathcal{N}$ of $\mathcal{X}$ generated by the closure $\mathcal{X}'$ of the commutator subgroup of $\mathcal{X}$ and the inertia groups I_{ik}. For this, let N be the intermediate field of the extension M_∞/L fixed by $\mathcal{N}$. Then N is the largest subfield of M_∞ which is abelian over L and unramified above p. Thus $M_0 \subset N$. If we assume that $M_0 \ne N$, we find an intermediate field N_0 of finite degree over L such that $M_0 \subsetneq N_0 \subset N$. Let $\mathfrak{N}$ be the conductor of N_0/L. Then the primes which divide $\mathfrak{N}$ are exactly the primes in T. Recall our definition $\mathfrak{M}_T = \prod_{\mathfrak{P} \in T} \mathfrak{P}$.
The commutative diagram

$$\begin{array}{ccccccc}
\mathfrak{o}_L & \longrightarrow & (\mathfrak{o}_L/\mathfrak{N})^\times & \longrightarrow & \mathrm{cl}_L^{\mathfrak{N}} & \twoheadrightarrow & \mathrm{cl}_L \\
\| & & \downarrow & & \downarrow & & \| \\
\mathfrak{o}_L & \longrightarrow & (\mathfrak{o}_L/\mathfrak{M}_T)^\times & \longrightarrow & \mathrm{cl}_L^T & \twoheadrightarrow & \mathrm{cl}_L
\end{array}$$

now implies that the order of the kernel of the surjection $\mathrm{cl}_L^{\mathfrak{N}} \twoheadrightarrow \mathrm{cl}_L^T$ is prime to p, since the only occurring primes are below the primes in T. What we have shown is $N_0 \subset M_0$, in contradiction to our assumption.

Lemma 3.2.8 *Let $\mathcal{X}'$ be the closure of the commutator subgroup of $\mathcal{X}$. Then*

$$\mathcal{X}' = X_T^{\gamma_L - 1}.$$

PROOF. The proof of Lemma 13.14 in [Wa] nearly remains unchanged. We only have to replace the inertia subgroup I_1 in loc.cit. by $\overline{\langle\gamma\rangle}$. □

Since $\gamma^j = \gamma$, the above Lemma implies that we get an isomorphism on minus parts

$$A_L^T \otimes \mathbb{Z}_p \simeq \left(X_T \overline{\langle\gamma\rangle} / \langle X_T^{\gamma_L - 1}, I_{ik}\rangle\right)^- \simeq X_T^- / \langle (X_T^-)^{\gamma_L - 1}, a_{ik}\rangle.$$

As already mentioned, the inertia groups I_{ik} are conjugate for fixed i, hence $\sigma_{ik} \equiv \sigma_{i1} \bmod \mathcal{X}'$ and likewise $a_{ij} \equiv a_{i1} \bmod \mathcal{X}'$ for all k. Hence

$$A_L^T \otimes \mathbb{Z}_p \simeq X_T^- / \langle (X_T^-)^{\gamma_L - 1}, a_1, \ldots, a_s\rangle,$$

where we have defined $a_i := a_{i1}$. Since $X_T^- / (X_T^-)^{\gamma_L - 1} = (X_T^-)_{\Gamma_L}$, Proposition 3.2.7 follows from the following lemma.

Lemma 3.2.9 *If $\mathfrak{P}_j = \mathfrak{P}_i^g$ for an element $g \in G$, then $a_j \equiv a_i^g \bmod (X_T^-)^{\gamma_L - 1}$.*

PROOF. Let $\tau \in \mathrm{Gal}(M_\infty/K)$ be a lift of g. Then g acts on $(X_T^-)_{\Gamma_L}$ via conjugation by τ. $\tilde{\mathfrak{P}}_{i1}^\tau$ is a prime in M_∞ above $\mathfrak{P}_j$, hence there exists an $x \in \mathcal{X}$ such that $\tilde{\mathfrak{P}}_{i1}^\tau = \tilde{\mathfrak{P}}_{j1}^x$. Replacing τ by $x^{-1}\tau$ we may assume that $x = 1$. Hence

$$\overline{\langle\sigma_{j1}\rangle} = I_{j1} = I_{i1}^\tau = \overline{\langle\sigma_{i1}^\tau\rangle}.$$

Since the restriction to L_∞ induces an isomorphism $I_{j1} \simeq \Gamma_L^{p^{n_j}}$ and

$$\sigma_{i1}^\tau|_{L_\infty} = (\gamma_L^{p^{n_i}})^\tau = (\gamma_L^{p^{n_i}})^g = \gamma_L^{p^{n_i}},$$

we have $n_i = n_j$ and $\sigma_{j1} = \sigma_{i1}^\tau$, i.e.

$$a_j = (a_i \gamma^{p^{n_j}})^\tau \cdot \gamma^{-p^{n_j}}.$$

But $\gamma^\tau|_{L_\infty} = \gamma_L$ implies that $\gamma^\tau = x_\tau \cdot \gamma$ for an element $x_\tau \in X_T$. We even have $x_\tau \in X_T^+$, since j trivially acts on γ and commutes with τ. Hence, the assertion follows from the above equation by taking minus parts. □

Chapter 4

On the Rubin-Stark conjecture

D. Burns [B3] has shown that the LRNC implies certain congruences of abelian L-functions at $s = 0$. These congruences in turn imply, among other things, the Rubin-Stark conjecture. We will reprove this result for the case at hand by a different method.

4.1 The conjecture

Let L/K be a finite abelian extension of number fields with Galois group G. Let S be a finite G-invariant set of primes of L, containing all the infinite primes and all the primes which ramify in L/K. If T is a second G-invariant, finite, nonempty set of primes of L, disjoint from S, we define for each character χ of G a complex-analytic function $\delta_T(\chi, s) = \prod_{\mathfrak{P} \in T^*}(1 - N(\mathfrak{p})^{1-s}\chi(\phi_{\mathfrak{P}}))$. The (S,T)-modified L-function associated to χ is defined to be

$$L_{S,T}(L/K, \chi, s) = \delta_T(\chi, s) \cdot L_S(L/K, \chi, s).$$

Set $\delta_T(s) = \sum_{\chi \in \mathrm{Irr}\,(G)} \delta_T(\check{\chi}, s)\varepsilon_\chi$ for all $s \in \mathbb{C}$. The S-Stickelberger and respectively (S,T)-Stickelberger functions[1] are defined by

$$\Theta_S(s) = \Theta_S(L/K, s) := \sum_{\chi \in \mathrm{Irr}\,(G)} L_S(L/K, \check{\chi}, s)\varepsilon_\chi,$$

$$\Theta_{S,T}(s) = \Theta_{S,T}(L/K, s) := \delta_T(s) \cdot \Theta_S(s) = \sum_{\chi \in \mathrm{Irr}\,(G)} L_{S,T}(L/K, \check{\chi}, s)\varepsilon_\chi.$$

We now fix a set of data $(L/K, S, T, r)$, where $r \geq 0$ is an integer, and which satisfies the following hypotheses (H):

- S contains all the infinite primes of L and all primes of L which ramify in L/K.
- S^* contains at least r primes which split completely in L/K.

[1] Do not confuse with the representing homomorphism Θ_S^T defined in 2.2.4.

- $|S^*| \geq r+1$.
- $T \neq \emptyset$, $S \cap T = \emptyset$, $E_S^T \cap \mu_L = 1$.

Since the common vanishing order $r_S(\chi)$ of $L_S(L/K, \chi, s)$ and $L_{S,T}(L/K, \chi, s)$ at $s = 0$ is at least r by [Ta2], Proposition 3.4, p. 24, we may define

$$\Theta_{S,T}^{(r)}(0) := \lim_{s \to 0} s^{-r} \Theta_{S,T}(s) \in \mathbb{C}G.$$

Here we think of $\Theta_{S,T}(s)$ as a holomorphic function in $s = 0$. Note that $\Theta_{S,T}^{(0)}(0) = \theta_S^T$ if $j \in G_{\mathfrak{P}}$ for all $\mathfrak{P} \in S$.

Now let us choose an r-tuple $W = (\mathfrak{P}_1, \ldots, \mathfrak{P}_r)$ of r distinct primes of S^* which split completely in L/K. We denote the r-th exterior power of a $\mathbb{Z}G$-module M by $\wedge_G^r M$. One defines a regulator map

$$\begin{array}{ccc} \mathbb{C} \wedge_G^r E_S^T & \xrightarrow{R_W} & \mathbb{C}G \\ e_1 \wedge \ldots \wedge e_r & \mapsto & \det\limits_{1 \leq i,j \leq r} \left(-\sum\limits_{g \in G} \log |e_j|_{\mathfrak{P}_i^g} g \right) \end{array}$$

for $e_1, \ldots, e_r \in E_S^T$, and then extending by $\mathbb{C}$-linearity. If R is a subring of $\mathbb{C}$ and M an RG-module without R-torsion, we define

$$M_{r,S} = \{x \in M | x \cdot \varepsilon_\chi = 0 \in \mathbb{C}M \ \forall \chi \in \mathrm{Irr}\,(G) \text{ such that } r_S(\chi) > r\}.$$

As proved in [Ru], R_W is a $\mathbb{C}G$-morphism, which induces an isomorphism

$$(\mathbb{C} \wedge_G^r E_S^T)_{r,S} \xrightarrow{\simeq} \mathbb{C}G_{r,S}.$$

For each $\Phi = (\phi_1, \ldots, \phi_{r-1}) \in (\mathrm{Hom}_{\mathbb{Z}G}(E_S^T, \mathbb{Z}G))^{r-1}$ one can define a $\mathbb{C}G$-morphism

$$\wedge\Phi : \mathbb{C} \wedge_G^r E_S^T \to \mathbb{C}E_S^T,$$

such that for all $e_1, \ldots, e_r \in \mathbb{C}E_S^T$ one has

$$\wedge\Phi(e_1 \wedge \ldots \wedge e_r) = \prod_{k=1}^{r} (-1)^{k+1} \det_{\substack{1 \leq i \leq r-1 \\ 1 \leq j \leq r,\ j \neq k}} (\phi_i(e_j)) \cdot e_k.$$

One defines a $\mathbb{Z}G$-submodule of $\mathbb{Q} \wedge_G^r E_S^T$ by

$$\Lambda_{S,T} = \begin{cases} \{\epsilon \in (\mathbb{Q} \wedge_G^r E_S^T)_{r,S} | \wedge \Phi(\epsilon) \in E_S^T \ \forall \Phi \in (\mathrm{Hom}_G(E_S^T, \mathbb{Z}G))^{r-1}\}, & r \geq 1 \\ \mathbb{Z}G_{0,S}, & r = 0. \end{cases}$$

We are now ready to state the Rubin-Stark conjecture as formulated by Rubin [Ru].

Conjecture 4.1.1 *Assume that the data $(L/K, S, T, r)$ satisfies (H). Then for any choice of W as above there exists a unique $\epsilon_{S,T,W} \in \Lambda_{S,T}$ such that $R_W(\epsilon_{S,T,W}) = \Theta_{S,T}^{(r)}(0)$.*

We will refer to this conjecture as $B(L/K,S,T,r)$. Note that the conjecture is independent of the choice of W, and that the uniqueness is automatic (cf. [P2], Remark 2 and 3). Further, $B(L/K,S,T,1)$ for varying S and T implies the Brumer-Stark conjecture as shown in [P2], Proposition 3.4.

Let p be a rational prime. If we replace $\Lambda_{S,T}$ by $\mathbb{Z}_{(p)}\Lambda_{S,T}$ in the above conjecture, we get a localized conjecture which we denote by $\mathbb{Z}_{(p)}B(L/K,S,T,r)$. One has

$$B(L/K,S,T,r) \iff \mathbb{Z}_{(p)}B(L/K,S,T,r)\ \forall p.$$

Our main tool in proving parts of the Rubin-Stark conjecture is the following theorem, which is Theorem 3.2.2.3 in [P3].

Theorem 4.1.2 *Assume that $(L/K,S,T,r)$ satisfies (H) and let $p \neq 2$ be a rational prime. Choose r distinct primes $\mathfrak{P}_1,\ldots,\mathfrak{P}_r \in S^*$ which split completely in L/K, and set $S_0 := S \setminus (G\mathfrak{P}_1 \cup \ldots \cup G\mathfrak{P}_r)$. Then it holds:*

$$\Theta_{S_0,T}(0) \in \mathrm{Fitt}_{\mathbb{Z}_pG_-}(A_L^T \otimes \mathbb{Z}_p) \Longrightarrow \mathbb{Z}_{(p)}B(L/K,S,T,r).$$

Moreover, we will need the following results which are taken from Proposition 2.3 in [P2].

Proposition 4.1.3 *Let p be a rational prime, and assume that the set of data $(L/K,S,T,r)$ satisfies (H). Then it holds:*

(1) If $S \subset S'$ and $(L/K,S',T,r)$ also satisfies (H), then

$$\mathbb{Z}_{(p)}B(L/K,S,T,r) \Longrightarrow \mathbb{Z}_{(p)}B(L/K,S',T,r)$$

(2) If $T \subset T'$ and $(L/K,S,T',r)$ also satisfies (H), then

$$\mathbb{Z}_{(p)}B(L/K,S,T,r) \Longrightarrow \mathbb{Z}_{(p)}B(L/K,S,T',r)$$

4.2 The tamely ramified case

We apply the results of the previous chapter to prove

Theorem 4.2.1 *Let L/K be an abelian Galois CM-extension with Galois group G and $p \neq 2$ a prime. Assume that for each prime $\mathfrak{p}$ above p the ramification is almost tame or $j \in G_{\mathfrak{p}}$. Then the minus part of the LRNC at p implies the Rubin-Stark conjecture $\mathbb{Z}_{(p)}B(L/K,S,T,r)$ for each sets of places S, T and each integer r such that $(L/K,S,T,r)$ satisfies (H).*

We immediately get from Theorem 3.2.2:

Corollary 4.2.2 *Assume that L/K additionally satisfies $j \in G_{\mathfrak{p}}$ for all primes $\mathfrak{p}$ above p, whenever $L^c \subset (L^c)^+(\zeta_p)$, and that $\mu_- = 0$. Then $\mathbb{Z}_{(p)}B(L/K,S,T,r)$ holds whenever $(L/K,S,T,r)$ satisfies (H).*

Note that we can again remove the condition $\mu_- = 0$ if $p \nmid |G|$.

Proof of Theorem 4.2.1. It follows from Theorem 4.1.2 and Proposition 4.1.3 that it suffices to show that $\Theta_{S_{\mathrm{ram}},T_0}(0) \in \mathrm{Fitt}_{\mathbb{Z}_pG_-}(A_L^{T_0} \otimes \mathbb{Z}_p)$ for minimal sets T_0. Hence, let $T_0 = \{\mathfrak{P}_0^g | g \in G\}$ for an unramified prime $\mathfrak{P}_0$ such that $E_{S_{\mathrm{ram}}}^{T_0} \cap \mu_L = 1$. This is equivalent to the statement on earlier occasions that $1 - \zeta \notin \prod_{g \in G/G_{\mathfrak{P}_0}} \mathfrak{P}_0^g$ for all $1 \neq \zeta \in \mu_L$. As before, define S_1 to be the set of all wildly ramified primes above p and set $T = T_0 \cup (S_{\mathrm{ram}} \setminus (S_{\mathrm{ram}} \cap S_p))$. By Theorem 2.3.1 the minus part of the LRNC at p implies (and is indeed equivalent to)

$$\mathrm{Fitt}_{\mathbb{Z}_pG_-}(A_L^T \otimes \mathbb{Z}_p) = (\theta_{S_1}^T). \tag{4.1}$$

We have two exact sequences

$$\left(\mathfrak{o}_L / \prod_{\mathfrak{P} \in T \setminus T_0} \mathfrak{P}\right)^{\times,-} \otimes \mathbb{Z}_p \rightarrowtail A_L^T \otimes \mathbb{Z}_p \twoheadrightarrow A_L^{T_0} \otimes \mathbb{Z}_p, \tag{4.2}$$

$$\left(\mathfrak{o}_L / \prod_{\mathfrak{P} \in T \setminus T_0} \mathfrak{P}\right)^{\times} \otimes \mathbb{Z}_p \rightarrowtail \bigoplus_{\mathfrak{P} \in T^* \setminus T_0^*} \mathrm{ind}_{G_\mathfrak{P}}^G T_\mathfrak{P} \otimes \mathbb{Z}_p \twoheadrightarrow \bigoplus_{\mathfrak{P} \in T^* \setminus T_0^*} \mathrm{ind}_{G_\mathfrak{P}}^G W_\mathfrak{P} \otimes \mathbb{Z}_p.$$

The first follows from sequence (2.8) for the sets T and T_0, whereas the second derives from diagram (2.12). We want to apply the following Lemma, which is a special case of Lemma 5 in [BG2].

Lemma 4.2.3 *Let $M_1 \rightarrowtail P_1 \to P_2 \twoheadrightarrow M_2$ be an exact sequence of finite $\mathbb{Z}_pG_-$-modules, where P_1 and P_2 are c.t. Then $\mathrm{Fitt}(P_i)$ is invertible for $i = 1, 2$ and*

$$\mathrm{Fitt}(M_2) = \mathrm{Fitt}(M_1^\vee) \cdot \mathrm{Fitt}(P_1)^{-1} \cdot \mathrm{Fitt}(P_2),$$

where $M_1^\vee = \mathrm{Hom}(M_1, \mathbb{Q}/\mathbb{Z})$ denotes the Pontryagin dual of M_1.

We have to modify the above two exact sequences slightly. For each prime $\mathfrak{P}$ we have an exact sequence

$$\mathcal{K}_\mathfrak{P} \rightarrowtail (\mathrm{ind}_{G_\mathfrak{P}}^G \mathbb{Z}_pG_\mathfrak{P}/(N(\mathfrak{P}) - 1))^- \twoheadrightarrow \left(\mathrm{ind}_{G_\mathfrak{P}}^G (\mathfrak{o}_L/\mathfrak{P})\right)^{\times,-} \otimes \mathbb{Z}_p,$$

where the second map is induced by mapping 1 to a generator of $(\mathfrak{o}_L/\mathfrak{P})^\times$. These sequences glue together and give

$$\mathcal{K} \rightarrowtail P \twoheadrightarrow \left(\mathfrak{o}_L / \prod_{\mathfrak{P} \in T \setminus T_0} \mathfrak{P}\right)^{\times,-} \otimes \mathbb{Z}_p, \tag{4.3}$$

where $\mathcal{K}$ and P are the direct sums of the $\mathcal{K}_\mathfrak{P}$ and the middle terms in the above sequence, respectively. Note that $\mathcal{K}$ and P are finite, and P is c.t. Define

$$c'_\mathfrak{P} := (|G_\mathfrak{P}|(1 - \frac{1}{|G_\mathfrak{P}|}N_{G_\mathfrak{P}}) + \frac{1}{|G_\mathfrak{P}|}N_{G_\mathfrak{P}}) \cdot c_\mathfrak{P} \in W_\mathfrak{P},$$

where $c_{\mathfrak{P}}$ was defined in (2.14). Moreover, let $t'_{\mathfrak{P}}$ be a preimage of $c'_{\mathfrak{P}}$ in $T_{\mathfrak{P}}$. The maps $\mathbb{Z}_pG_{\mathfrak{P}} \to W_{\mathfrak{P}} \otimes \mathbb{Z}_p$, $1 \mapsto c'_{\mathfrak{P}}$ and $\mathbb{Z}_pG_{\mathfrak{P}} \to T_{\mathfrak{P}} \otimes \mathbb{Z}_p$, $1 \mapsto t'_{\mathfrak{P}}$ are injective and become isomorphisms after tensoring with $\mathbb{Q}_p$. Hence, the direct sum

$$\mathcal{T} := \bigoplus_{\mathfrak{P}\in T^*\setminus T_0^*} \operatorname{ind}_{G_{\mathfrak{P}}}^{G} T_{\mathfrak{P}}/t'_{\mathfrak{P}} \otimes \mathbb{Z}_p$$

is finite and c.t. by Lemma 2.3.3. Therefore, the sequences (4.2) and (4.3) give two exact sequences

$$\mathcal{K} \rightarrowtail P \to A_L^T \otimes \mathbb{Z}_p \twoheadrightarrow A_L^{T_0} \otimes \mathbb{Z}_p,$$

$$\mathcal{K} \rightarrowtail P \to \mathcal{T}^- \twoheadrightarrow \mathcal{W}^-,$$

where $\mathcal{W}$ is the direct sum of the $\operatorname{ind}_{G_{\mathfrak{P}}}^{G} W_{\mathfrak{P}}/c'_{\mathfrak{P}} \otimes \mathbb{Z}_p$. We can apply Lemma 4.2.3 to these sequences and get

$$\operatorname{Fitt}(A_L^{T_0} \otimes \mathbb{Z}_p) = \operatorname{Fitt}(A_L^T \otimes \mathbb{Z}_p) \cdot \operatorname{Fitt}(\mathcal{T}^-)^{-1} \cdot \operatorname{Fitt}(\mathcal{W}^-). \tag{4.4}$$

Proposition 2.3.5 (4) implies

$$\operatorname{Fitt}(\mathcal{T}^-) = \prod_{\mathfrak{P}\in T^*\setminus T_0^*} (\tau_{\mathfrak{P}}), \tag{4.5}$$

$$\tau_{\mathfrak{P}} = f_{\mathfrak{P}}(1-q_{\mathfrak{p}})\frac{1}{|G_{\mathfrak{P}}|}N_{G_{\mathfrak{P}}} + (|G_{\mathfrak{P}}| - N_{G_{\mathfrak{P}}})(\frac{q_{\mathfrak{p}} - \phi_{\mathfrak{P}}}{1-\phi_{\mathfrak{P}}}\varepsilon_{\mathfrak{P}} + 1 - \varepsilon_{\mathfrak{P}}),$$

where as before $\varepsilon_{\mathfrak{P}} = |I_{\mathfrak{P}}|^{-1}N_{I_{\mathfrak{P}}}$, $q_{\mathfrak{p}} = N(\mathfrak{p})$, and $f_{\mathfrak{p}}$ is the degree of the corresponding residue field extension.

Lemma 4.2.4 *Let $\mathfrak{P} \notin S_p$ be a finite prime of L. Then*

$$\operatorname{Fitt}_{\mathbb{Z}_pG_{\mathfrak{P}}}(W_{\mathfrak{P}}/c'_{\mathfrak{P}} \otimes \mathbb{Z}_p) = \langle N_{G_{\mathfrak{P}}} - |G_{\mathfrak{P}}|, N_{G_{\mathfrak{P}}} + (N_{I_{\mathfrak{P}}} - f_{\mathfrak{P}}^{-1}N_{G_{\mathfrak{P}}})(\phi_{\mathfrak{P}} - 1)^{-1}\rangle_{\mathbb{Z}_pG_{\mathfrak{P}}}.$$

PROOF. Since $\mathfrak{P}$ lies not above p, we may assume that $\mathfrak{P}$ is at most tamely ramified. Keep the notation of Lemma 2.0.14. Define a map

$$\pi : \mathbb{Z}G_{\mathfrak{P}}e_1 \oplus \mathbb{Z}G_{\mathfrak{P}}e_2 \to W_{\mathfrak{P}}$$

by $\pi(e_1) = (b^{-1}-1, 1)$ and $\pi(e_2) = (a-1, 0)$. We claim that the kernel is generated by $N_{I_{\mathfrak{P}}}e_2$ and $(a-1)e_1 + (1-b^{-1})e_2$. For this, assume that

$$\pi(x_1e_1 + x_2e_2) = (x_1(b^{-1}-1) + x_2(a-1), \overline{x}_1) = 0 \in W_{\mathfrak{P}}.$$

By Lemma 6.6 in [Ch2] $x_1 = (a-1)x'_1$ for an appropriate $x'_1 \in \mathbb{Z}G_{\mathfrak{P}}$. By the same Lemma in loc.cit. we get $x'_1(b^{-1}-1) + x_2 = y \cdot N_{I_{\mathfrak{P}}}$ for a $y \in \mathbb{Z}G_{\mathfrak{P}}$, since the left hand side is annihilated by $(a-1)$. This proves the claim.
Define two group elements

$$\delta_1 := \sum_{i=0}^{f_{\mathfrak{P}}-1} b^{-i} + (b^{-1}-1)^{-1}(1 - |b|^{-1}\sum_{i=0}^{|b|-1} b^i) \in \mathbb{Z}_pG_{\mathfrak{P}},$$

$$\delta_2 := \sum_{i=0}^{e_{\mathfrak{P}}-c_{\mathfrak{P}}-1} a^i - \sum_{j=0}^{c_{\mathfrak{P}}^{-1}e_{\mathfrak{P}}-1} \left(\sum_{k=0}^{c_{\mathfrak{P}}-1} k a^{c_{\mathfrak{P}} j + k} \right) \cdot \sum_{i=0}^{f_{\mathfrak{P}}-1} b^i \in \mathbb{Z}_p G_{\mathfrak{P}}.$$

An easy but lengthy computation shows that $\pi(\delta_1 e_1 - \delta_2 e_2) = c'_{\mathfrak{P}}$. Hence, the kernel of the epimorphism

$$\mathbb{Z}_p G_{\mathfrak{P}} e_1 \oplus \mathbb{Z}_p G_{\mathfrak{P}} e_2 \twoheadrightarrow W_{\mathfrak{P}}/c'_{\mathfrak{P}} \otimes \mathbb{Z}_p$$

induced by π is generated by the kernel of π and $\delta_1 e_1 - \delta_2 e_2$. From this one can compute the desired Fitting ideal. □

Recall the definitions (2.7) and (2.10) of ω and the modules $M_{\mathfrak{P}}$. The above Lemma together with (4.4), (4.1), (4.5) now yields

Corollary 4.2.5

$$\mathrm{Fitt}_{\mathbb{Z}_p G_-}(A_L^{T_0} \otimes \mathbb{Z}_p) = (q_{\mathfrak{p}_0} - \phi_{\mathfrak{P}_0})\omega \prod_{\mathfrak{P} \in S^*_{\mathrm{ram}}} M_{\mathfrak{P}} \subset SKu(L/K)^- \cdot \mathbb{Z}_p G.$$

In particular, this implies

$$\Theta_{S_{\mathrm{ram}},T_0}(0) = (q_{\mathfrak{p}_0} - \phi_{\mathfrak{P}_0}) \cdot \omega \prod_{\mathfrak{P} \in S^*_{\mathrm{ram}}} (1 - \varepsilon_{\mathfrak{P}} \phi_{\mathfrak{P}}^{-1}) \in \mathrm{Fitt}_{\mathbb{Z}_p G_-}(A_L^{T_0} \otimes \mathbb{Z}_p),$$

which proves Theorem 4.2.1. □

REMARK. As one can see from the results in [GK], it is not true in general that $\Theta_{S_{\mathrm{ram}},T_0}(0)$ lies in the Fitting ideal of $A_L^{T_0} \otimes \mathbb{Z}_p$. But note that all the counterexamples in loc.cit. are wildly ramified above p. Thus, we have actually shown a stronger result (which is called the Strong Brumer-Stark Conjecture in [P3]).

Appendix A

Removing $\mu_- = 0$

We combine methods used by J. Ritter and A. Weiss [RW5], A. Wiles [Wi1] and C. Greither [Gr1] to remove the hypothesis $\mu_- = 0$ in Theorem 3.1.2 (2) for a special class of cases, including the case $p \nmid |G|$. More precisely, we prove

Theorem A.0.6 *Let T be the set of places of L defined in (3.1). Suppose that for each prime $\mathfrak{p} \in T(K)$ at least one of the following conditions is satisfied:*

- $j \in I_{\mathfrak{p}}$
- $j \notin I_{\mathfrak{p}}$, *but* $j \in G_{\mathfrak{p}}$ *and* $N(\mathfrak{p})^{f_{\mathfrak{p}}/2} \not\equiv -1 \bmod p$
- $p \nmid |I_{\mathfrak{p}}|$

Then we have

$$\mathrm{Fitt}_{\mathbb{Z}_p[[\mathcal{G}]]_-}(X_T^-) = (\Psi_T).$$

REMARK. In the proof of Theorem 3.2.2 we have enlarged the extension L/K to L'/K. But if L/K satisfies the hypotheses of the above theorem, then so does L'/K.

PROOF. Since the projective dimension of X_T^- as a $\mathbb{Z}_p[[\mathcal{G}]]_-$-module is at most 1 by Proposition 3.1.1, the Fitting ideal in demand is principal, generated by $\tilde{\Psi}_T$, say. The integral closure of $\mathbb{Z}_p[[\mathcal{G}]]_-$ is $R := \sum_\chi \mathbb{Z}_p[\chi][[T]]$, where the sum runs over all odd irreducible characters of $\tilde{G}$. Since $\mathbb{Z}_p[[\mathcal{G}]]_- \cap R^\times = (\mathbb{Z}_p[[\mathcal{G}]]_-)^\times$, it suffices to show

(1) $R\tilde{\Psi}_T = R\Psi_T$

(2) $(\tilde{\Psi}_T) \subset (\Psi_T)$.

If χ is an odd irreducible character of $\tilde{G}$ and X is any $\mathbb{Z}_p[[\mathcal{G}]]_-$-module, we define $\mathbb{Z}_p[\chi][[T]]$-modules

$$X_\chi := X \otimes_{\mathbb{Z}_p[[\mathcal{G}]]_-} \mathbb{Z}_p[\chi][[T]],$$

$$\begin{aligned} X^\chi &:= \left\{ x \in \mathbb{Z}_p[\chi] \otimes_{\mathbb{Z}_p} X \,|\, gx = \chi(g)x \ \forall g \in \tilde{G} \right\} \\ &= \mathrm{Hom}_{\mathbb{Z}_p[\chi]\tilde{G}}(\mathbb{Z}_p[\chi], \mathbb{Z}_p[\chi] \otimes_{\mathbb{Z}_p} X). \end{aligned}$$

To prove (1) we have to show that $\mathrm{Fitt}_{\mathbb{Z}_p[\chi][[T]]}((X_T^-)_\chi)$ is generated by $\chi(\Psi_T)$. By (1) of Theorem 3.1.2 this holds apart from the μ-invariants. By Lemma 3.3 in [Gr1] there is an isomorphism $(X_T^-)_\chi \simeq X_T^\chi$, since X_T^- is c.t. over $\tilde{G}$. Moreover, the epimorphism $X_T^- \twoheadrightarrow X_{\mathrm{std}}^-$ has a kernel C which is finitely generated as $\mathbb{Z}_p$-module (cf. (3.2)), and thus it induces an exact sequence

$$C^\chi \rightarrowtail X_T^\chi \to X_{\mathrm{std}}^\chi \twoheadrightarrow H^1(\tilde{G}, \mathrm{Hom}_{\mathbb{Z}_p[\chi]}(\mathbb{Z}_p[\chi], \mathbb{Z}_p[\chi] \otimes_{\mathbb{Z}_p} C)),$$

where the rightmost term is finite. Hence, $\mu(X_T^\chi) = \mu(X_{\mathrm{std}}^\chi)$, and the latter equals the μ-invariant of $\chi(\Psi_T)$ by Theorem 1.4 in [Wi1] if χ is of order prime to p. For the general result one has to adjust the (second part of the) proof of Theorem 16 in [RW5]. As already mentioned earlier, one should think of the claim of Theorem A.0.6 as a reformulation of the equivariant Iwasawa main conjecture; hence equation (1) states that the conjecture is true over the maximal order R, which is Theorem 6 in [RW3].

It remains to prove (2). Write $\tilde{G} = G' \times \tilde{G}_p$, where $\tilde{G}_p$ is the p-Sylow subgroup of $\tilde{G}$, and thus $p \nmid |G'|$. We have a natural decomposition

$$\mathbb{Z}_p[[\mathcal{G}]]_- = \bigoplus_{\substack{\chi' \in \mathrm{Irr}\,(G') \\ \chi' \,\mathrm{odd}}} R(\chi'),$$

where $R(\chi') = \mathbb{Z}_p[\chi'][[\tilde{G}_p \times \Gamma_K]]$ is a local ring. Its maximal ideal $\mathfrak{m}_{\chi'}$ is generated by p and the augmentation ideal $\Delta[[\tilde{G}_p \times \Gamma_K]]$. We define a prime ideal $P_{\chi'} := (p, \Delta\tilde{G}_p) \subsetneq \mathfrak{m}_{\chi'}$.

Lemma A.0.7 *For each $\mathfrak{p} \in T(K)$ the element $\xi_\mathfrak{p}$ defined in (3.4) becomes a unit in $R(\chi')_{P_{\chi'}}$.*

PROOF. Recall the definition $Z_\mathfrak{p} = \mathrm{ind}_{\mathcal{G}_\mathfrak{p}}^{\mathcal{G}} \mathbb{Z}_p$. As one can learn from the proof of Proposition 8 in [Gr2], we have

$$(\xi_\mathfrak{p}) = \mathrm{Fitt}_{\mathbb{Q}_p[[\mathcal{G}]]_-}(\mathbb{Q}_p Z_\mathfrak{p}(1)^-)\mathrm{Fitt}_{\mathbb{Q}_p[[\mathcal{G}]]_-}(\mathbb{Q}_p Z_\mathfrak{p}^-)^{-1}.$$

But $Z_\mathfrak{p}^- = 0$ if $j \in G_\mathfrak{p}$. Moreover, $Z_\mathfrak{p}(1) = \mathbb{Z}_p[[\mathcal{G}]]/\langle q_\mathfrak{p} - \phi_\mathfrak{p}, \tau - 1, \tau \in I_\mathfrak{p} \rangle$, where as before $q_\mathfrak{p} = N(\mathfrak{p})$. Hence, $Z_\mathfrak{p}(1)^- = 0$ if $j \in I_\mathfrak{p}$. Now assume that $j \notin I_\mathfrak{p}$, but $j \in G_\mathfrak{p}$ and $q_\mathfrak{p}^{f_\mathfrak{p}/2} \not\equiv -1 \bmod p$. Then $\phi_\mathfrak{p}^{f_\mathfrak{p}/2} - q_\mathfrak{p}^{f_\mathfrak{p}/2} \equiv j - q_\mathfrak{p}^{f_\mathfrak{p}/2} \bmod T$, and $j - q_\mathfrak{p}^{f_\mathfrak{p}/2}$ becomes a unit on minus parts. This means that $\phi_\mathfrak{p}^{f_\mathfrak{p}/2} - q_\mathfrak{p}^{f_\mathfrak{p}/2}$ is a unit in $\mathbb{Z}_p[[\mathcal{G}]]_-$, and hence $Z_\mathfrak{p}(1)^- = 0$ in this case, too. We have proven so far that $\xi_\mathfrak{p}$ is actually a unit in $\mathbb{Z}_p[[\mathcal{G}]]_-$ if $\mathfrak{p}$ satisfies the first or the second condition of the theorem. We are left with the case $p \nmid |I_\mathfrak{p}|$.

It suffices to show that $(1-\phi_{\mathfrak{p}}^{-1}q_{\mathfrak{p}})\varepsilon_{\mathfrak{p}}+1-\varepsilon_{\mathfrak{p}}$ and $(1-\phi_{\mathfrak{p}})\varepsilon_{\mathfrak{p}}+1-\varepsilon_{\mathfrak{p}}$ become units at $P_{\chi'}$. We only treat the first element, the other case is similar.
For this, we have to prove that $\chi'((1-\phi_{\mathfrak{p}}^{-1}q_{\mathfrak{p}})\varepsilon_{\mathfrak{p}}+1-\varepsilon_{\mathfrak{p}}) \notin P_{\chi'}$. Assume that this is false. Since $1 \notin P_{\chi'}$, we must have $\chi'(\varepsilon_{\mathfrak{p}}) = 1$. Let us write $\phi_{\mathfrak{p}}^{-1} = \sigma' \cdot \sigma_p \cdot \gamma_K^c$, where $\sigma' \in G'$, $\sigma_p \in \tilde{G}_p$, $0 \neq c \in \mathbb{Z}_p$. Since $\sigma_p - 1 \in P_{\chi'}$, we have $1-\chi'(\sigma')\gamma_K^c q_{\mathfrak{p}} = 1-\chi'(\sigma')q_{\mathfrak{p}}(1-T)^c \in P_{\chi'}$. Since $P_{\chi'}$ contains no unit, we must have $p|(1-\chi'(\sigma')q_{\mathfrak{p}})$, and hence $1-(1-T)^c \in P_{\chi'}$. If we write $c = p^n \cdot \alpha$, $\alpha \in \mathbb{Z}_p^\times$, we find out that $1-(1-T^{p^n})^\alpha \in P_{\chi'}$. Finally, $1-(1-T^{p^n})^\alpha = T^{p^n} \cdot g(T)$ with a power series $g(T)$ with $g(0) = -\alpha$, hence $g(T)$ is a unit. This implies $T \in P_{\chi'}$, a contradiction. □

We now return to the proof of Theorem A.0.6. The epimorphism $X_T^- \twoheadrightarrow X_{\mathrm{std}}^-$ implies the first inclusion in

$$\mathrm{Fitt}_{R(\chi')}((X_T^-)_{\chi'}) \subset \mathrm{Fitt}_{R(\chi')}((X_{\mathrm{std}}^-)_{\chi'}) \subset (G_{(\chi')^{-1}\omega, S_{\mathrm{ram}} \cup S_p}(T)),$$

whereas the second inclusion is (10), p. 562 in [Wi2]. Localizing at $P_{\chi'}$ gives

$$(\chi'(\tilde{\Psi}_T))_{P_{\chi'}} \subset (G_{(\chi')^{-1}\omega, S_{\mathrm{ram}} \cup S_p}(T))_{P_{\chi'}} = (\chi'(\Psi_T))_{P_{\chi'}},$$

since all the $\xi_{\mathfrak{p}}$ become units at $P_{\chi'}$. Therefore, there is an element $r' \in R(\chi') \setminus P_{\chi'}$ such that $r' \cdot \chi'(\tilde{\Psi}_T) \in (\chi'(\Psi_T))$. We already know from Theorem 3.1.2 that one can find a positive integer i such that $p^i \cdot \chi'(\tilde{\Psi}_T) \in (\chi'(\Psi_T))$. Hence

$$(p^i, r')(\chi'(\tilde{\Psi}_T)) \subset (\chi'(\Psi_T))$$

and the ideal (p^i, r') has finite index in $R(\chi')$.
Thus, $(\chi'(\tilde{\Psi}_T)) + (\chi'(\Psi_T))/(\chi'(\Psi_T))$ is a submodule of $R(\chi')/(\chi'(\Psi_T))$ of finite cardinality. Now the proof following (10.5) in [Wi1] shows that the only such module is trivial. We obtain $(\chi'(\tilde{\Psi}_T)) \subset (\chi'(\Psi_T))$, and thus we get (2). This completes the proof of the theorem. □

Bibliography

[BrB] Breuning, M., Burns, D.: *Leading terms of Artin L-functions at $s = 0$ and $s = 1$* Compos. Math. **143**, No. 6 (2007), 1427-1464

[Bl] Bley, W.: *On the equivariant Tamagawa number conjecture for abelian extensions of a quadratic imaginary field*, Documenta Math. **11** (2006), 73-118

[BB] Bley, W., Burns, D.: *Equivariant epsilon constants, discriminants and étale cohomology*, Proc. London Math. Soc. **87** (2003), 545-590

[B1] Burns, D.: *Equivariant Tamagawa numbers and Galois module theory I*, Compos. Math. **129**, No. 2 (2001), 203-237

[B2] Burns, D.: *Equivariant Whitehead Torsion and Refined Euler Characteristics*, CRM Proceedings and Lecture Notes **36** (2003), 35-59

[B3] Burns, D.: *Congruences between derivates of abelian L-functions at $s = 0$*, Invent. Math. **169** (2007), 451-499

[BG1] Burns, D., Greither, C.: *On the equivariant Tamagawa number conjecture for Tate motives*, Invent. Math. **153** (2003), 305-359

[BG2] Burns, D., Greither, C.: *Equivariant Weierstrass preperation and values of L-functions at negative integers*, Doc. Math. Extra Vol. (2003), 157-185

[Ca] Cassou-Noguès, P.: *Valeurs aux entiers négatifs des fonctions zêta et fonctions zêta p-adiques*, Invent. Math. **51** (1979), 29-59

[Ch1] Chinburg, T.: *On the Galois structure of algebraic integers and S-units*, Invent. Math. **74** (1983), 321-349

[Ch2] Chinburg, T.: *Exact sequences and Galois module structure*, Ann. Math. **121** (1985), 351-376

[CR2] Curtis, C. W., Reiner, I.: *Methods of Representation Theory with applications to finite groups and orders*, Vol. 2, John Wiley & Sons, (1987)

[FW] Ferrero, B., Washington, L.: *The Iwasawa invariant μ_p vanishes for abelian number fields*, Ann. Math. **109** (1979), 377-395

[Fl] Flach, M.: *The equivariant Tamagawa number conjecture: a survey.* in Burns, D., Popescu, C., Sands, J., Solomon, D. (eds.): *Stark's Conjectures: Recent work and new directions*, Papers from the international conference on Stark's Conjectures and related topics, Johns Hopkins University, Baltimore, August 5-9, 2002, Contemporary Math. **358** (2002), 79-125

[Gr1] Greither, C.: *Some cases of Brumer's conjecture for abelian CM extensions of totally real fields*, Math. Zeitschrift **233** (2000), 515-534

[Gr2] Greither, C.: *Computing Fitting ideals of Iwasawa modules*, Math. Zeitschrift **246** (2004), 733-767

[Gr3] Greither, C.: *Determining Fitting ideals of minus class groups via the Equivariant Tamagawa Number Conjecture*, to appear in Compos. Math.

[GK] Greither, C., Kurihara, M.: *Stickelberger elements, Fitting ideals of class groups of CM fields, and dualisation*, to appear in Math. Zeitschrift

[GRW] Gruenberg, K. W., Ritter, J., Weiss, A.: *A Local Approach to Chinburg´s Root Number Conjecture*, Proc. London Math. Soc. (3) **79** (1999), 47-80

[GW] Gruenberg, K. W., Weiss, A.: *Galois invariants of local units*, Quart. J. Math. Oxford **47** (1996), 25-39

[Ha] Hayes, D.R.: *Stickelberger functions for non-abelian Galois extensions of global fields* in Burns, D., Popescu, C., Sands, J., Solomon, D. (eds.): *Stark's Conjectures: Recent work and new directions*, Papers from the international conference on Stark's Conjectures and related topics, Johns Hopkins University, Baltimore, August 5-9, 2002, Contemporary Math. **358** (2002), 193-206

[Ku] Kurihara, M.: *Iwasawa theory and Fitting ideals*, J. Reine Angew. Math. **561** (2003), 39-86

[Neu] Neukirch, J.: *Algebraische Zahlentheorie*, Springer (1992)

[NSW] Neukirch, J., Schmidt, A., Wingberg, K.: *Cohomology of number fields*, Springer (2000)

[RW1] Ritter, J., Weiss, A.: *A Tate sequence for global units*, Compos. Math. **102** (1996), 147-178

[RW2] Ritter, J., Weiss, A. : *The Lifted Root Number Conjecture and Iwasawa Theory*, Memoirs of the AMS **748** (2002)

[RW3] Ritter, J., Weiss, A. : *Toward equivariant Iwasawa theory*, Manuscripta Mathematica **109** (2002), 131-146

[RW4] Ritter, J., Weiss, A. : *Representing* $\Omega_{(l\infty)}$ *for real abelian fields*, J. of Algebra and its Appl. **2** (2003), 237-276

[RW5] Ritter, J., Weiss, A. : *Toward equivariant Iwasawa theory, II*, Indagationes Mathematicae **15** (2004), 549-572

[Ta1] Tate, J.: *The cohomology groups of tori in finite Galois extensions of number fields*, Nagoya Math. J. **27** (1966), 709-719

[Ta2] Tate, J.: *Les conjectures de Stark sur les fonctions L d'Artin en* $s = 0$, Birkhäuser, (1984)

[P1] Popescu, C.D.: *Base change for Stark-type conjetures "over* $\mathbb{Z}$*"*, J. Reine Angew. Math. **542** (2002), 85-111

[P2] Popescu, C.D.: *On the Rubin-Stark conjecture for a special class of CM-extensions of totally real number fields*, Math. Zeitschrift **247**, No. 3 (2004), 529-547

[P3] Popescu, C.D.: *Rubin's Integral Refinement of the Abelian Stark Conjecture*, in Burns, D., Popescu, C., Sands, J., Solomon, D. (eds.): *Stark's Conjectures: Recent work and new directions*, Papers from the international conference on Stark's Conjectures and related topics, Johns Hopkins University, Baltimore, August 5-9, 2002, Contemporary Math. **358** (2002), 1-35

[Ru] Rubin, K.: *A Stark conjecture "over* $\mathbb{Z}$*" for abelian L-functions with multiple zeros*, Annales de L'Institut Fourier **46** (1996), 33-62

[Wa] Washington, L. C.: *Introduction to Cyclotomic Fields*, Springer (1982)

[We] Weiss, A.: *Multiplicative Galois module structure*, Fields Institute Monographs 5, American Mathematical Society (1996)

[Wi1] Wiles, A.: *The Iwasawa conjecture for totally real fields*, Ann. Math. **131**, 493-540 (1990)

[Wi2] Wiles, A.: *On a conjecture of Brumer*, Ann. Math. **131**, 555-565 (1990)

Index

admissible isomorphism, 23
Artin conductor, 57
Artin root number, 5, 41, 57

character
    cyclotomic —, 80
    even, 53
    odd, 53
    symplectic —, 41, 45
    Teichmüller —, 81
CM-extension, 51
complex conjugation, 51
Conjecture
    Equivariant Tamagawa Number —, 6
    Iwasawa main —, 8, 81, 82, 98
    Lifted Root Number —, 5, 41, 84
    — for small S, 46
    $\mu = 0$, 8, 82, 97
    Rubin-Stark —, 6, 8, 92, 93
    Stark —, 41, 45, 59
    Strong Brumer-Stark —, 8, 96
cyclotomic extension, 79

Dirichlet map, 17, 51
    modified —, 41, 43
Dual, 9

Fitting ideal, 63, 82, 84, 87, 94
functional equation, 57

Hom description, 15, 53

inertial lattice, 16, 41–42
Iwasawa algebra, 80
Iwasawa module, 80, 82
    standard —, 80, 82
Iwasawa series, 81

L-function, 40
    (S,T)-modified —, 91
    completed Artin —, 57
    p-adic —, 81
Localization sequence, 10, 82

Ω_ϕ, 18

ray class group, 54, 64, 85
refined Euler characteristic, 28
regulator, 58
regulator map, 92
relative class number, 57

Section, 9
    commutative —, 10
Sinnott-Kurihara ideal, 62, 96
Stark-Tate regulator, 40
    modified —, 43, 78
Stickelberger element, 59, 60, 84, 96
Stickelberger function, 91

Tate twist, 80
Tate-sequence, 6
    — for small S, 7, 15
    construction of —, 20–23
transport, 25

variance
    — with S, 30, 36
    — with ϕ, 28

LEBENSLAUF

Geburtstag:	19.03.1979
Geburtsort:	Augsburg
Familienstand:	verheiratet mit Frau Maren Nickel, geb. Möller seit dem 18.05.2007
Kinder:	1 Sohn, Malte Hjalmar Nickel, geb. am 09.05.2008

1985-89:	Besuch der Grundschule am Eichenwald, Neusäß
1989-98:	Besuch des humanistischen Gymnasiums bei St. Stephan, Augsburg
1998:	Abitur
1998-99:	Zivildienst im Krankenhauszweckverband Augsburg
1999-2004:	Studium der Mathematik an der Universität Augsburg
2004:	Erlangung des akademischen Grades eines Diplom-Mathematikers an der Universität Augsburg
2005-08:	Wissenschaftlicher Mitarbeiter und Promotionsstudent an der Universität Augsburg